Embryology, Epigenesis, and Evolution

Historically, philosophers of biology have tended to sidestep the problem of development by focusing primarily on evolutionary biology and, more recently, on molecular biology and genetics. Quite often, development has been misunderstood as simply, or even primarily, a matter of gene activation and regulation. Nowadays a growing number of philosophers of science are focusing their analyses on the complexities of development; in *Embryology, Epigenesis, and Evolution*, Jason Scott Robert explores the nature of development against current trends in biological theory and practice and looks at the interrelations between evolution and development (evo–devo), an area of resurgent biological interest.

Clearly written, this book should be of interest to students and professionals in the philosophy of science and the philosophy of biology.

Jason Scott Robert is Assistant Professor of Philosophy at Dalhousie University and Canadian Institutes of Health Research New Investigator.

Embryology, Epigenesis, and Evolution

Taking Development Seriously

JASON SCOTT ROBERT

Dalhousie University

CAMBRIDGE UNIVERSITY PRESS
Cambridge, New York, Melbourne, Madrid, Cape Town, Singapore, São Paulo

Cambridge University Press
The Edinburgh Building, Cambridge CB2 2RU, UK

Published in the United States of America by Cambridge University Press, New York

www.cambridge.org
Information on this title: www.cambridge.org/9780521824675

First published 2004
This digitally printed first paperback version 2006

A catalogue record for this publication is available from the British Library

Library of Congress Cataloguing in Publication data

Robert, Jason Scott.
    Embryology, epigenesis, and evolution : taking development seriously / Jason Scott Robert.
      p. cm. – (Cambridge studies in philosophy and biology)
    Includes bibliographical references and index.
    ISBN 0-521-82467-2
    1. Developmental biology – Philosophy. 2. Embryology – Philosophy. 3. Evolution
    (Biology) – Philosophy. I. Title. II. Series.
    QH491.R63   2004
    571.8 – dc21   2003048461

ISBN-13 978-0-521-82467-5 hardback
ISBN-10 0-521-82467-2 hardback

ISBN-13 978-0-521-03086-1 paperback
ISBN-10 0-521-03086-2 paperback

Contents

List of Figures

Preface

Developmental biology, as a science, is full of mystique. My imagination was easily captured by the amazing journey of the organism from egg to adult, the robustness of development under variable conditions, and the remarkable emergence of complexity during ontogeny. The project of understanding development is one taken up by philosophers of biology only recently, despite having its roots in Aristotle, and I suspect one of the reasons is that development has always been shrouded in a tapestry woven of vitalistic strands so long anathema to philosophers working on the natural sciences.

In the main, philosophers of biology have tended to sidestep development; they have, instead, tended to analyse the apparently more tractable problems of evolutionary biology (particularly fitness) and, more recently, molecular biology and genetics (particularly the relation between classical and molecular genetics). There are, of course, exceptions to these tendencies. In fact, a growing number of philosophers of biology are now exploring the complexities of development. This book catalogues some of the most interesting aspects of this philosophical work, in the context of a sustained introduction to recent developmental science and theory.

In the following pages, I offer a philosophical account of organismal development, address the character of developmental mechanisms, and argue that we should resist the assumption that development can be explained exclusively in terms of gene action and activation. Drawing on this account, I also engage a series of biological and conceptual issues in understanding the relationship between development and evolution – another area of substantial recent philosophical interest.

It is nice – no, necessary – to use real examples in the philosophy of biology. I use them throughout this book. And yet, sometimes, fictional examples can be heuristically useful. A case in point is *Jurassic Park*. Readers (and viewers) of *Jurassic Park* may be forgiven for believing that many of the problems of the development of organisms have been solved. To be sure, things go wrong for the dinosaur makers – plenty of things go wrong. Nevertheless, they succeed in building dinosaurs, however monstrous, from DNA. Deoxyribonucleic acid, we have come to understand, is the key component of life. DNA distinguishes between life and non-life, and once we have isolated the DNA, so the story goes, we can understand all life forms, extant and extinct.

That is an overarching dream of genetics and genomics, and, although DNA, in concert with ecology and development, became a nightmare for the dinosaur makers in *Jurassic Park*, the dream persists in the minds and work of many biologists. Consider the efforts to map and sequence all the genes in a hypothetical, abstract human genome, that of Hugo. (The name 'Hugo' derives from that of one of the Human Genome Project's international overseers, the Human Genome Organization.) Hugo's genome is *hypothetical* because it is not the genome of any particular human, but rather is assembled, in a patchwork manner, from the genomes of hundreds of humans worldwide. (That said, the genome sequenced by Celera Genomics independent of the publicly funded Human Genome Project is, in fact, largely the genome of Celera's former president, Craig Venter; see Wade 2002.) Nonetheless, these genomes are *abstract* because they are said to be representative of humans in general, although that is a physical impossibility, not least because we know of no core sequence of nucleotides (not even Venter's) shared by all humans whatever.

Nevertheless, the first working drafts of a complete human genome were published in *Nature* (International Human Genome Sequencing Consortium 2001) and *Science* (Venter et al. 2001) in February 2001 by the Human Genome Project and by Celera Genomics. With the draft sequences now at hand, scientists are proceeding to finalise the draft(s) and are beginning to annotate and understand the human genome. However, there are deep conceptual problems involving the relationship between Hugo's genome and the development of a human organism. The most glaring ones involve his genome's strikingly artificial nature: genomes simply do not exist independent of the complex organisms of which they are but one part; organisms are not genomes writ large.

Consider, for instance, whatever happened to development. Concerns abound regarding the relationship of genotypes to phenotypes: how does Hugo's gerrymandered, pristine genome correspond to the vast genetic diversity discernible in humans? Is the genome the prime ingredient in a human being, or are there other such ingredients? More basically, is there even any such thing as a 'prime' ingredient in an organism? Is the genome itself sufficient to produce a human animal? If not, what else is required besides DNA to yield a complex, specifically structured, functional human? What are the steps and processes, twists and turns, leading from gametic through geriatric humanity?

As we enter the post-genomic era, the real work is just beginning, for there is a vast developmental terrain to traverse between a genome sequence and a complex, functional organism – as the dinosaur makers in *Jurassic Park* learned all too well.

There are epistemological, metaphysical, and methodological impediments to Hugo's graceful coming out. Many of us have, by necessity, begun to take development into account. Genes don't work by themselves; they must be made to work in developmental context, and they must be reproduced through the generations. A standard interpretation is that the inherited genome initiates and directs development, and that we can understand the development of organisms best by beginning with the genome and investigating the minutiae of gene activation. I contend that this interpretation is misguided, that there is much more to development than the activation of genes, and that the genome may be the wrong place to start in understanding development.

Taking development into account is not the same as taking development seriously.[1] To take development seriously is not to hide behind metaphors of the magical powers of genes – they 'instruct' or 'program' the future organism. To take development seriously is rather to explore in detail the processes and mechanisms of differentiation, morphogenesis, and growth, and the actual (not ideologically or perhaps merely technologically inflated) roles of genes in these organismal activities. Despite the existence of what has come to be known as the 'interactionist consensus', according to which everyone agrees that both genes and environments 'interact' in the generation (and explanation) of organismal traits, my claim is that those swept up in genomania have nonetheless failed to take development seriously.

This is a book about the philosophy of developmental biology in relation to genetics and genomics, and so also in relation to evolutionary biology. This is not, however, a book about gene (or genome) bashing. I take the critical role of DNA in development seriously, but my primary *explanandum* is development, as set within the epistemological and methodological contexts of genomics and genetics.

Developmental biology has played a curious role in the biological mainstream of the past century or so. The halcyon days of evolutionary and experimental embryology eventually gave way to the experiments of the classical geneticists; with the synthesis of a number of biological subdisciplines in the 1940s under the aegis of population genetics, embryology had virtually no presence; when embryology was later reborn as developmental biology, it had virtually no choice but to be guided in the main by the new molecular genetics. Only lately have biologists begun to reconsider how development may not be fully explained in terms of differential gene expression, and how evolutionary biology, relatively ignorant of development, can provide only an incomplete account of the nature and processes of evolution. Drawing on a selective history of key themes in embryology and developmental biology, the burden of this book is to motivate a more integrative approach to biology and to provide a conceptual framework for understanding the central place of development in biology.

I have chosen an alliterative title, *Embryology, Epigenesis, and Evolution*, to describe the investigation in these pages. In Chapter 1, I explore what it might mean to take development seriously, in theory and in practice. I begin by detailing the central problem of development: how it is that a relatively simple, homogeneous cellular mass can become a relatively complex, heterogeneous organism. Then, framed within a discussion of some heuristics employed in developmental biology and of the impact of the respective biases of these heuristics, I critically assess the use made of certain kinds of experiments in supporting and preserving the overwhelming sense that development can be explained strictly or primarily in terms of differential gene expression. In Chapter 2, I orient the reader with three examples, one each to represent the three elements of the book's title: embryology (the experiments of Roux and Driesch); epigenesis (homeobox genes in development and evolution); and evolution (blind cave fish).

In Chapter 3, I begin to worry about particular metaphors commonly associated with the explanation of development in recent years – 'genetic programme', 'genetic information', 'triggering', and 'interaction' – and I explore their impact on biological theory and practice. Aspects of the old debate between preformationists and epigenecists comprise the subject matter of Chapters 3 and 4. Historically, preformationists held that a future individual organism is somehow contained in toto in the zygote or, more ambitiously, in either the ovum or the sperm (depending on one's sex-cell preference). The future organism merely grows into a fully formed adult without an attendant increase in complexity. Epigenecists held, to the contrary, that the complex individual, guided by some directing principle, emerges from relative homogeneity over developmental time; the future organism is formed during ontogenesis rather than pre-existing it. There are neither pure preformationists nor pure epigenecists in the world today; in fact, most views of development seek to meld aspects of preformationism with elements of epigenesis. I examine various of these modern reconciliations, representing what

I take to be the 'modern consensus' view on development, before exploring in subsequent chapters the problems with this position.

In Chapter 4, I briefly recount salient elements of the history of the conceptual and professional divorce between genetics and embryology achieved in the early part of the twentieth century. This potted history should not be taken as either authoritative or comprehensive; its function is, rather, illustrative, and it motivates an extended discussion of 'epigenetics' and an argument for the desirability of organism-centred biology. I also offer my own account of epigenetics as constitutive of genes, which serves as the basis for my discussion in Chapter 5 of the creativity of development.

In the final two chapters, I turn squarely to the third element of my title: evolution. Chapter 6 details the most promising synthesis of development, genetics, and evolution to date – evolutionary developmental biology (evo–devo). I provide a series of examples to show how development and evolution, and developmental and evolutionary explanations, can be interrelated, and I argue that taking development into account may well offer a substantive challenge to evolutionary theory as we know it. Then, in Chapter 7, I return to the question of taking development seriously. There I explore conceptual and theoretical aspects of the relationship between evo–devo and the developmental systems perspective on ontogenetic processes in evolution, indicating the benefits and limits of both approaches and elucidating the fallout of taking development seriously. Again, the aim is not to belittle the role of genes in development and evolution but rather to establish a clear and realistic sense of what genes can and cannot do for us. When we take development seriously, I contend, it becomes apparent that the explanatory burden is not discharged at the level of genes in either developmental or evolutionary contexts.

Enormous debts of gratitude are owed to those who have tried to set me right. Among the scholars who read part or all of various drafts of this manuscript in various forms, or who engaged me in particular debates along the way, and whose comments are deeply appreciated, are Barry Allen, Rich Campbell, Ford Doolittle, Gill Gass, Russell Gray, Jim Griesemer, Paul Griffiths, Brian Hall, Evelyn Fox Keller, Manfred Laubichler, Alan Love, Wendy Olson, Susan Oyama, Bob Perlman, Rudolf Raff, Michael Ruse, Roger Sansom, Sahotra Sarkar, Ken Schaffner, Kim Sterelny, Jon Stone, Rob Wilson, Bill Wimsatt, and several anonymous referees.

Audiences at meetings of the American Philosophical Association (Central and Eastern Divisions), the Atlantic Region Philosophers Association, the Canadian Philosophical Association, the Canadian Society for History and

Philosophy of Science, and the International Society for History, Philosophy, and Social Studies of Biology provided very helpful feedback, as did those who heard me speak at Dalhousie University (my home department), Duke University, McGill University, the University of Calgary, the University of Texas at Austin, and the University of Western Ontario. Special thanks are due to members of the Philosophy and Developmental Biology Group, particularly Dick Burian, Werner Callebaut, Scott Gilbert, and Lenny Moss.

The figures were redrawn by Tim Fedak, for which I thank him; I also thank the relevant publishers for their permission to reproduce copyrighted material. Chapter 6 appeared in slightly modified form in *Biology & Philosophy* 17, 591–611 (2002).

The Fulbright Foundation, the Social Sciences and Humanities Research Council of Canada, the Killam Trust of Dalhousie University, and the Canadian Institutes of Health Research (and the CIHR Institute of Genetics) provided generous research funding along the way.

To my parents, Judi and John, my sister, Keitha, and my partner, Wanda, I owe the greatest debts: for perseverance, love, and endless support. This book is dedicated to them.

1

The Problem of Development

It is not good enough to answer [questions regarding development] by saying it is simply a matter of turning some genes on and others off at the right times. It is true that molecular biology provides numerous detailed precedents for mechanisms by which this can, in principle, be done, but we demand something more than these absolutely true, absolutely vacuous statements.

– *Sydney Brenner* (1974)

The central problem of developmental biology is to understand how a relatively simple and homogeneous cellular mass can differentiate into a relatively complex and heterogeneous organism closely resembling its progenitor(s) in relevant respects. This is not a new problem. It has been with us since Aristotle, at least. However, it is only recently that we have established a handle on how possibly to solve it. I am not convinced that we have yet grasped the right handle, though.

A decade ago, an advertisement for *The Encylopedia of the Mouse Genome* appeared in a biotechnology serial. The tagline read: 'The Complete Mouse (some assembly required)' (cited in Gilbert and Faber 1996: 136). The parenthetical clause refers, of course, to development. As those of us who have purchased ready-to-assemble furniture know all too well, this is indeed an onerous requirement, for the assembly process may very well have the greatest impact on final outcome! What is true of ready-to-assemble furniture is also true, I contend, of organisms believed to be 'ready-to-assemble' from DNA and assorted other material.

No one honestly believes that development can be achieved unilaterally by genes acting alone or in concert. Rather, everyone agrees that genes are important to, but not sufficient for, development. This is so, ontogenetically at

1

least (and perhaps also ontologically, for those concerned with ontology), and serves as the basis for the recent 'interactionist consensus' on development: the view that neither genes nor environments, neither nature nor nurture, suffices for the production of phenotypes.

I want to take this further: genes are important to, but not sufficient for, not only development but also the *explanation* of development. This epistemic and methodological claim is more controversial than the ontogenetic truism at the core of the interactionist consensus. My burden is to diminish the controversy surrounding this claim, in part by unpacking the interactive assembly of organisms.

In this chapter, my strategy is to explore a number of methodological principles used in biology; the first two of them are general, and the next three are used specifically in the context of understanding development. I provide arguments, abstracted from the biological and philosophical literature, for both the use of heuristics as such (the first principle) and for the use of particular heuristics (the second principle). For rhetorical purposes, I interpret the five principles as premises in an argument aimed at explaining development. I then illustrate how variance in the interpretation and application of the second principle yields inconsistent results and biases our biological knowledge in various ways. I argue in favour of an unorthodox reading of one of the heuristics, but a reading required by the imperative to take development seriously. In the chapters that follow, I further explore this imperative.

HEURISTICS

It is fair to say that biological phenomena are a messy lot. Though this may often be true in other domains as well, in biology, at least, a staggering number of simplifying assumptions must be made just to get a research programme off the ground. Historically, the most significant simplifying assumptions (or heuristics) employed in genetics and developmental biology have resulted in the elision of the organism as both nexus and nadir of developmental interactions. For the most part, these heuristics are well justified; they are, at least, widely accepted. Nevertheless, differences in how they are interpreted and applied generate differences in what we can claim to know about development.

Let us define 'heuristics' as *simplifying strategies to be used in situations of cumbersome investigational complexity* (Wimsatt 1980, 1986c; Gigerenzer et al. 1999). One crucial caveat about heuristics is that they are purpose relative. As Wimsatt notes, 'all instruments in the natural, biological and social sciences are designed for use in certain contexts and can produce biased or worthless results if they are used in contexts that may fail to meet the conditions for which they were designed' (Wimsatt 1986c: 297). Examples might include the use of analysis of variance as a surrogate for the analysis of causes (Lewontin 1974; Sober 2000); the application of the methods of quantitative genetics where the assumptions of quantitative genetics (linearity, additivity, constancy, and so on) do not hold (Pigliucci and Schlichting 1997); or the use of linkage analysis in psychiatric genetics where the conditions of successful linkage (single gene of major effect, clear diagnostic criteria, known pattern of inheritance, and clinical homogeneity amongst affected family members) are not met (Robert 2000a). In using heuristics, then, we must be careful to select the right one(s).

That notwithstanding, without the use of heuristics, we would be much further from solutions to pressing biological problems than we currently are. Here, then, is a universally acknowledged premise of biological research:

1. Simplifying strategies and assumptions, as such, are absolutely necessary in biological science.

This is an heuristic dealing with the use of reductionistic heuristics. There are at least twenty reductionistic heuristics in widespread use today, including those used in conceptualisation, model building, theory construction, experimental design, observation, and interpretation; Wimsatt has documented these heuristics, and also their characteristic biases (Wimsatt 1980, 1986c).

Unlike Laplacian demons, human investigators of all stripes have limited intellectual, computational, temporal, and financial capacities. Any biological system to be studied must be simplified in various ways to make it tractable for agents like us. The very reason that we build simplified models is that we are limited beings, and most of the systems we want to understand are too complex in their natural state; thus we abstract from them what seem to be the most important or the most easily manipulated variables in order to generate a manageable representation of their workings.

One of the most common heuristic strategies is to simplify the *context* of a system under study. If we want to learn about *intrasystemic* causal factors – that is, if we want to learn about what's going on inside a particular system – we build a model or design an experiment wherein the context of the system is simplified rather than the system itself. Of course, we sometimes have to do both, especially if the system of interest is particularly complex; in such

3

a case, we might use another kind of reductionistic strategy. But a golden rule of experimental design is this: simplify the context first. Hence, a second general principle of biological methodology:

2.   Simplifying the context of a system is advantageous if we want to learn about intrasystemic causal factors.

Amongst those who hold to the interactionist consensus, the strategy of context simplification is extensively employed in investigations of the role of genes in development, usually in the form of 'environmental control'. Here, one holds environmental variables constant across experiments or, worse, actually believes that the environment simply is invariant. One standard approach is to vary genetic factors against a common, invariant background of environmental factors – a standard environment. Context simplification, instantiated as environmental control, is the basic methodological framework of many researchers creating and employing genome sequence data, for instance. Sequence data are produced by isolating strands of DNA, cloning them, and employing a variety of techniques to ascertain the order of nucleotides and their physical relationship to each other. Genomes, or even individual strands of DNA – the systems under study – do not exist in isolation from natural environments except in the pristine artificiality of the lab; moreover, as we shall see in later chapters, there are good reasons to believe that even the structure (let alone the functions) of strands of DNA cannot be understood in isolation from their organismal context. Nevertheless, the environments, broadly construed, of DNA were abstracted away and held constant in the effort to generate the sequence of the human genome. (The same is true, of course, of the genome sequences of model organisms, such as the mouse and the nematode worm.) The context was simplified, the experimental work proceeded, and draft versions of the genome sequence are now at hand.

For the most part, and despite occasional slips to the contrary, biologists are careful in employing the strategy of context simplification. For instance, with rare but notable exceptions – such as Hamer and Copeland (1998), but see Hamer (2002) – very few scientists or commentators would today suggest that either nature (genes) or nurture (environments) is singularly decisive in organismal development. Despite the standard use of experimental or interpretive techniques to partition causation into internal (natural, genetic) and external (nurturing, environmental) components, techniques which may be unable by their very design to detect interactions between genes and environments (Wahlsten 1990; Sarkar 1998), most scholars grant that phenotypic traits arise from complex, possibly nonadditive, interactions between multiple factors at many hierarchical levels.

However, not all varieties of interactionism are equivalent, and a vigorous debate has arisen over which varieties in fact take interaction seriously, and which simply pay 'lip service' to interaction in a reflexive refrain masking secret adherence to the old nature–nurture debate (Robert 2003). This debate will figure prominently in the paragraphs that follow, as well as in later chapters in the discussion of how best to interpret the second premise.

EXPLORING DEVELOPMENT

Let me now briefly spell out three additional premises, again universally granted, which are employed as additional steps, beginning with the first two premises, in (roughly) a chain of argument putatively leading to a conclusion about development.

The third premise, already alluded to, states the following:

3.   Genes by themselves are not causally efficacious, as genes and environments (at many scales) interact (differentially, over time) in the generation of any phenotypic trait.

Whereas, once upon a time, biologists and commentators may have been happy to claim that genes determine organisms, body and mind alike, just as other scientists (mainly social scientists) and commentators were happy to claim that the organism is a kind of tabula rasa to be inscribed, shaped, and structured entirely by experience, no one seriously (or, at least, no one justifiably) entertains either of those perspectives today. It is for this reason that scientists are happy to declare the nature–nurture debate dead, settled in favour of both (Goldsmith et al. 1997). There are no (overt) genetic determinists these days, even though some environmental determinists persist (usually in an effort to ward off the spectre of genetic determinism). As Russell Gray has put it, 'nowadays it seems that everybody is an "interactionist"' (Gray 1992: 172). So much so, in fact, that those perceived to be stirring the ashes of the nature–nurture debate are called nasty names and relegated to the periphery of accepted scientific practice. This is the legacy of the interactionist consensus.

The fourth premise is designed to permit investigation of interacting variables in development (in line with premises 1 and 2):

4.   We decide to focus on the causal agency of genes against a constant background of other factors, for pragmatic or heuristic reasons.

Experimental tractability is a core scientific *desideratum*. It is nice to imagine the world as full of interconnected parts not meaningfully separable from each other; but just try to analyse the world so imagined and science grinds to a halt. It turns out that genes are much more experimentally tractable than a wide range of other interacting factors and agents. This may be, of course, simply because we have spent so many decades perfecting techniques for genetic manipulation, and that huge amounts of money are available for such activities compared with others (Griffiths and Knight 1998: 255; Robert 2001b). Given the enormous amount of money available to study gene sequences, it is little wonder that genetic manipulation is quite easy compared with the experimental manipulation of other factors in development.

Nevertheless, it is worth briefly describing two scientifically well-regarded philosophical analyses justifying premise 4, such that premise 4 is universally acknowledged. First, Schaffner has published a careful study of the role of genes in the behavioural development of the nematode worm, *Caenorhabditis elegans*. Though he (and the scientists he studies) is well aware that genes must be coupled with other molecules within an organism in order to be causally efficacious (premise 3), Schaffner contends (in line with premises 1 and 2, and in support of premise 4) that '*epistemically* and *heuristically*, genes do seem to have a *primus intra pares* status'. This is in part because 'methods have been developed to screen for mutants, map "genes for" traits (as a first approximation), localise those genes, clone them, and test their role as "necessary" elements for a trait using sophisticated molecular deletion and rescue techniques' (Schaffner 1998: 234). With such methods in place, not starting with genes seems methodologically foolhardy. The embryologist Ross Harrison aptly noted early in the twentieth century that 'the investigator enters where he can gain a foothold by whatever means may be available' (Harrison 1918; cited by Gilbert and Sarkar 2000: 4).

A second, and related, justification for premise 4 is laid out by Gannett. She has analysed how genes come to be identified as causes primarily for pragmatic reasons (Gannett 1999). Having ruled out as unsuccessful the efforts of those who attempt to apply objective criteria (namely, causal priority, nonstandardness, and causal efficacy) to single out genes as causes, she argues that practical, and not theoretical, considerations are at play. Drawing on the work of Collingwood and van Fraassen on the context dependence of causal explanations, Gannett shows that what we identify as 'the' cause, amongst competing, equally necessary causes, depends jointly on the capacity to manipulate it (scientists' 'handle' – or, in Harrison's term, their 'foothold') and also the specific purposes of investigators (what sorts of questions are found meaningful and worthy of attention).

Pragmatic factors structure both of these contingencies: the capacity for manipulation is a function of past choices in, for instance, the development of particular technologies, and the questions found meaningful are decided by investigative aims, the practical end sought – for instance, the treatment or prevention of disease. Both contingencies are also deeply influenced by the availability of research funds; with the Human Genome Project, countless lab scientists suddenly saw a need for expensive gene-sequencing machines. Gannett concludes that, given the (necessary) incompleteness of causal explanations, whatever causal explanation offered will be both partial and pragmatically determined.

What we identify as a cause has its causal effects only in combination with additional necessary conditions (which, for other pragmatic reasons, might have themselves been identified as causes). This idea is epitomised in a fifth and final premise, one that may seem more controversial than the first four but is nonetheless widely acknowledged:

5. A trait x is caused by a gene y only against a constant background of supporting factors (conditions), without which x would not be present (even if y is present).

Prima facie, given premise 2, this fifth premise is a close relative of premise 3. Variations on this fifth premise have been employed as definitions of a 'genetic trait'. Consider Sterelny and Kitcher's sophisticated treatment:

> An allele A at a locus L in a species S is for trait P^* (assumed to be a determinate form of the determinable characteristic P) relative to a local allele B and an environment E just in case (a) L affects the form of P in S, (b) E is a standard environment, and (c) in E organisms that are AB have phenotype P^*. (Sterelny and Kitcher 1988: 350)

In other words, as long as that particular allele, in genetic and standard environmental context, is associated with the relevant phenotypic outcome, then that particular allele may be deemed an 'allele for' that phenotype. Given the necessity of simplifying assumptions (premises 1 and 2), as long as we recognise the critical contextual qualifications (premise 3) and also that we focus on allele A for heuristic and pragmatic reasons (premise 4), then we may deem premise 5 to be a plausible singling out of a gene as a cause in organismal development. So far, so good.

To reiterate, the five premises we have before us are as follows:

1. Simplifying strategies and assumptions, as such, are absolutely necessary in biological science.

2. Simplifying the context of a system is advantageous if we want to learn about intrasystemic causal factors.
3. Genes by themselves are not causally efficacious, as genes and environments (at many scales) interact (differentially, over time) in the generation of any phenotypic trait.
4. We decide to focus on the causal agency of genes against a constant background of other factors, for pragmatic or heuristic reasons.
5. A trait x is caused by a gene y only against a constant background of supporting factors (conditions), without which x would not be present (even if y is present).

These five premises taken together are usually thought to justify the following conclusion:

6. Therefore, organismal development is a matter of gene action and activation, as particular alleles have their specific phenotypic effects against standard environmental background conditions.

This conclusion coheres nicely with the standard explanation for why organisms develop as they do: there is a programme or set of instructions for development inscribed in the genes. Of course, genes alone do not an organism make. The genetic program must be activated or 'triggered', as there is no unmoved mover in the world as we know it; and the DNA must be suitably housed in appropriate cellular and extracellular contexts, which may themselves be very complex, in order for development to proceed. However, given these caveats, the specificity of development – the reliable, transgenerational reconstruction of form – is widely held to be best explained as a matter of gene action and activation.

But is that in fact true? Is development in fact *explained* in terms of gene action and activation? My argument is that it is not, though we all happily agree, at least in the abstract, with the five premises thought to generate it. Are we then illogical or, worse, illogical because we are ideologically motivated? Or is it rather the case that the five universally acknowledged premises do not actually generate the inference to the usual conclusion? I interpret the inference to the orthodox conclusion as invalid: the conclusion does not follow from the premises we have before us, because there are two mutually exclusive possible readings of the second premise just detailed, only one of which could be taken to support the conclusion. (Even were the second premise perfectly straightforward, as it does, indeed, seem to be, and even were we therefore justified in asserting the conclusion on the basis of the five premises, we would be mistaken to interpret the conclusion as specifying an *explanation of development* – a point to which I return in later paragraphs.)

Recall that premise 2 stipulates that simplifying the context of a system is advantageous if we want to learn about intrasystemic causal factors. Context simplification is usually achieved by holding certain factors constant while solving for others, and decisions about what to hold constant and what to investigate are pragmatically motivated, as already explained. However, the pragmatic dimension of these decisions renders the second premise crucially ambiguous: what counts as a system is not a matter of objective determination but is itself influenced by pragmatic factors, such that what counts as intrasystemic or extrasystemic is decided by a range of considerations and not, as it were, thrust at us by nature. Accordingly, our results are constrained by the experimental design and not the facts of nature.

Several systematic problems (what Wimsatt calls 'biases') are associated with environmental control as a context simplifier. First, context simplification is biased toward lower explanatory levels, so simplifying the environmental context stems from, and leads to, focusing on simple components of a system. Higher-level components of systems, and higher-level systems, are legislated out of epistemological and methodological existence in favour of lower-level systems and their components. Consequently, an investigator who simplifies the context in line with premise 2 may well be guilty of simplificatory asymmetry (Wimsatt 1986c: 300, 301). Second, we may be prone, should we forget or fail to appreciate the gravity of the simplifying assumption, to draw unjustified causal inferences; it is remarkably easy to fall into the trap of generating causal stories about genes against a constant environmental background (which itself exists only in the laboratory) – hence our fifth premise. We must be eternally vigilant, in simplifying the context, not to exaggerate the conclusions we draw.

I suggested earlier that premise 5 strikes us as entirely justified by appeal to premises 1 through 4. However, there is no necessity in my particular formulation of premise 5, nor in Sterelny and Kitcher's instantiation of this premise. Consider that, by parity of reasoning, we might just as well have (again for some pragmatic reason) postulated not an 'allele for' P^* but rather an 'extracellular environment for' P^* given standard allelic, cytoplasmic, and other environmental contexts (Gray 1992; Smith 1992; Mahner and Bunge

1997; Robert 2000c). That we do not postulate such 'extracellular environments for' does not imply that they do not exist; it implies, rather, that we have decided, for whatever reasons, that 'alleles for' are more important to establish. We are thereby guilty of explanatory asymmetry inasmuch as we a priori construe the relevant system in strictly reductionistic terms, thereby inviting inference to the conclusion that development is a genetic affair.

This result is fostered by only one of the 2 possible interpretations of premise 2. Both interpretations are heuristics in their own right. I shall refer to the suspect one as the 'hedgeless hedge' heuristic (HHH); the other, to be explored and defended in later paragraphs, is the 'constant factor principle' heuristic.

The phrase 'hedgeless hedge' is attributed to Roger McCain, who diagnosed hedgeless hedging as a major limitation of early sociobiological thinking (McCain 1980; see also Neumann-Held 1999). The notion, though, is more broadly applicable than that. A typical definition of 'hedging' is protecting oneself from loss or failure by undertaking a counterbalancing action, as in hedging one's bets by not placing all one's eggs in a single basket (an awkward mixture of metaphors, to be sure!). Hedgeless hedging is a win–win strategy, denoting a fail-safe type of hedging: one puts virtually all one's faith in *A* and relatively little in *B* and then attempts to establish *A* but not *B*; but betting on *B* at all (say, by publicly announcing that *B* is true, likely, or possible) provides a measure of safety just in case *B* and not *A*. Less formally, in proceeding according to the HHH, 'one admits the existence of an anomaly or problem of theory and then proceeds as though one had not. If one is then accused of neglecting the anomaly, one then produces the admission of its existence as conclusive evidence of one's innocence of the charge' (McCain 1980: 126). The hedgeless hedge is well characterised as a simplifying assumption, in particular a simplification of context: one admits the implausibility of the simplifying assumption but proceeds with the simple model nonetheless, generating results inadequate to the reality of the situation; when challenged, one refers back to the original admission of implausibility for exoneration.

McCain's example of this strategy is sociobiologists' treatment of inheritance. Although complexes of many genes (polygenes) are involved in the generation of any trait, for purposes of tractability the early models of sociobiological inheritance – such as that advanced in E.O. Wilson's *Sociobiology: The New Synthesis* (Wilson 1975) – reverted to one-locus theory, according to which we assume that one and only one gene is associated with a given inherited trait. As Wilson's mathematical models depend so heavily on one-locus theory, and the assumption of single loci is so inadequate to the reality of both inheritance and development, the model is rendered immediately suspect. McCain observes that Wilson is well aware of his simplifying assumption, and Wilson notes that future models will have to take polygenism into consideration; but to take polygenism into consideration is so completely to undermine the model on which Wilson's treatment of sociobiology rests that the one-locus model itself is virtually worthless. Nevertheless, admitting the limitations of the model functions as a hedge against the probability that the model is in fact not at all a good one.

The HHH shares with all heuristics the property of fallibility, which is a function of the cost effectiveness of heuristic use. However, the failures of heuristics tend to be systematic rather than random, such that we might identify these failures and correct for them (often by applying a new heuristic). That is, thanks to the systematic biases of simple heuristics, we are able to learn from our false models in generating truer, more complex theories (Wimsatt 1987). What is unique about the hedgeless hedge is that the limitations of the heuristic are so obvious that, even though a hedgelessly hedged model may initiate the production of more adequate models, such models will themselves be so drastically different from the original model that its catalytic role may be overestimated. Moreover, the HHH wears its bias on its sleeve, implying that its putative openness is sufficient to make the heuristic appear honest and true. Unlike other context simplification heuristics, the HHH contains within itself the additional mechanism of theoretical exoneration, thereby providing an excuse for denying, say, complexity while nonetheless admitting the existence (and importance) of such complexity.

There are abundant examples of hedgeless hedging in biological research. Elisabeth Lloyd has explored a curious phenomenon, one that she refers to as 'ritual recitation' (my 'reflexive refrain'), whereby investigators favourably cite the papers of those who have challenged the investigators' theoretical framework, perhaps to demonstrate awareness of the ideas of detractors, but then proceed as if there are in fact no problems with the framework. According to Lloyd, there is 'a peculiar disconnect between what the authors explicitly acknowledge as serious theoretical and evidential problems, and how they actually theorize and evaluate evidence' (Lloyd 1999: 225).

In illustrating this claim, Lloyd discusses the emerging field of evolutionary psychology. According to Lloyd, central texts in evolutionary psychology are rife with footnotes citing, for instance, Gould and Lewontin's paper on the limits of adaptationism (Gould and Lewontin 1979), indicating awareness of problems of panadaptationist evolutionary theory, and sometimes acknowledging the need to avoid committing the errors Gould and Lewontin warn against. But, as Lloyd shows, these citations are smuggled into monographs expressly giving adaptation by natural selection an exclusive role in the evolutionary origin of phenotypic traits. Accused of naive adaptationism, the authors may simply point to the references as putative evidence of their innocence. The issue here, as elsewhere, is 'a matter of the *actual weight given in practice* – not in lip-service' to the *B* term of the HHH (Lloyd 1999: 226).[1]

HEDGING ABOUT THE HOMEOBOX

Ritual recitation as an instance of hedgeless hedging is evident in philosophical commentaries on biology as well as in actual biological practice. But philosophers tend to go beyond ritual recitation in their application of the hedgeless hedge, building more sophisticated safeguards into the heuristic. Consider Alex Rosenberg's use of this heuristic in his critical analysis of physicalist antireductionism. Rosenberg defines physicalist antireductionism as the coupling of two theses: 'physicalism – the thesis that biological systems are nothing but physical systems, with antireductionism – the thesis that the complete truth about biological systems cannot be told in terms of physical science alone' (Rosenberg 1997: 446). He identifies this sort of coupling as a consensus view amongst philosophers of biology, and he interprets recent findings in developmental molecular biology as a substantive challenge to physicalist antireductionism.

Following Lewis Wolpert, Rosenberg asserts that, from 'the total DNA sequence and the location of all proteins and RNA' (Wolpert 1994: 571), we could predict the development of an embryo or, alternatively, compute, or even construct, the embryo.[2] Of course, as will be demonstrated in the paragraphs that follow, genetic research does not aim at the study of development as such, but rather strictly at the role that genes play against a constant developmental background (van der Weele 1999: 24); but Rosenberg takes the additional, unwarranted step of interpreting the genetic research as providing a complete explanation of development.

Rosenberg is interested in a class of genes known as the 'homeobox genes'. Widely, though problematically, referred to as 'master genes', the homeobox genes are often interpreted as crucial developmental switches which 'trigger' large numbers of downstream genes in the generation of complex structures, such as eyes (Robert 2001a). Rosenberg asserts the 'computability' of the embryo from a small number of 'stock elements', particularly DNA, RNA, and proteins, as directed by members of the class of homeobox genes. To avoid triviality, Rosenberg places what he takes to be a necessary constraint on the computability claim, namely that a computable algorithm must not advert to any cellular structures 'not themselves "computable" from the nucleic acids and the proteins that compose the fertilized egg' (Rosenberg 1997: 450). That said, he asserts that the essence of developmental molecular biology is to assume certain constant factors (e.g., inherited cellular structures and environmental context) and then to explain the whole of development 'without adverting further to ineliminable cellular physiology' (p. 455).

In defending this claim, Rosenberg ritually recites what he takes to be a truism, namely that 'the molecular developmental biologist cannot simply build an eye, still less an animal *in vitro*, by combining the right macromolecules in the right proportions in the right sequence, in the right intervals' because 'the cellular milieu in which these reactions take place is causally indispensable' (Rosenberg 1997: 454). Rosenberg thus subscribes to the interactionist consensus. So committed, he proceeds to interpret the role of the *Eyeless* gene and its homologues in eye morphogenesis against such a supportive background. He makes the claim:

> one of the most complex of organs is built by the switching on of a relatively small number of the same genes, across a wide variety of species, and that the great differences between, say mammalian eyes, and insect eyes, are the result of a relatively small number of regulatory differences in the sequence and quantities in which the same gene products are produced by genes all relatively close together on the chromosome, and that these genes build the eye without the intervention of specialized cellular structures beyond those required for any developmental process. Identifying the other genes in the cascade that produces the entire eye should in principle be a piece of normal science, which will enable the developmental geneticist to 'compute' the eye from nucleic acids and proteins alone. For if switching on *Eyeless* can create the eye, surely its creation is 'computable' at least in principle. (Rosenberg 1997: 454)

As there is no room in this story for causal explanations above the molecular level, physicalist antireductionism falters.

The in-principle computability of the embryo from a description of DNA, RNA, and proteins (the A term in the HHH) is by definition set against a constant background of supporting factors (the B term). If challenged, Rosenberg may point to his admission of their importance as evidence that he is guilt free. Rosenberg hedges here by defining core elements of the constant background, notably cell structures and activities, as themselves computable in the same way the rest of the embryo is. He does this in order to avoid triviality, as already noted. However, if we grant him this move – and we should not – then his conclusion follows necessarily.

13

Rosenberg attempts to defend the controversial move in two ways. First, he asserts that 'cellular structures only come into existence through the molecular processes that precede them. There is in developmental molecular biology therefore no scope for claims about the indispensable role of cellular structures in these molecular processes. The future cannot cause the past' (Rosenberg 1997: 455). Of course, no one is claiming that causation works against the arrow of time; but even if molecular processes do indeed occur before (and concurrently with, and after) cellular processes, it is an open question whether cellular processes and structures are in fact explicable (or even predictable[3]) from a description of molecular processes and structures. Rosenberg forecloses the question by sleight of hand in requiring that cellular structures be computable; momentarily, I will show that this foreclosure is suspect.

Rosenberg's second strategy is to claim that the very possibility of ever explaining development turns on the particular features of the computability claim he endorses: 'unless the vast diversity of form is . . . explainable from a tractable base of a relatively small number of regulatory and structural genes (and their protein products) combined by a similarly small number of combination rules, *we can surrender all hope* of any completeness and generality in the [sic] understanding how diversity in development is possible, let alone actual' (Rosenberg 1997: 451). Thus either we succumb to Rosenberg's conclusion or give up on understanding development altogether.

Most developmental biologists would, with justice, take issue with this putative dilemma. Developmental biologists almost uniformly hold that development is hierarchical, characterised by the emergence of structures and processes not entirely predictable (let alone explicable) from lower-level (e.g., genetic) properties of the embryo. A leading example of the fact that the development of an organism is not fully prescribed in its inherited zygotic or maternal DNA is cellular behaviour during morphogenesis. Despite Rosenberg's admonitions, cells' collective behaviour during morphogenesis simply cannot be either predicted or explained by examining the behaviour of individual cells (or, for that matter, DNA) prior to cell division, differentiation, or condensation (Hall and Miyake 1992, 1995, 2000; Hall 1999, 2000a). This is because the formation of cell condensations is contingent not on the directives of some imagined genetic programme but rather on the spatiotemporal state of the organism and its constituent modules. Developmental biologists, therefore, hold to a kind of physicalist antireductionism, offering the methodological advice that we must engage in multileveled investigation of ontogeny in order not to miss key features at microlevels, mesolevels, and macrolevels. Moreover, and again despite Rosenberg's admonitions, these biologists *qua* physicalist antireductionists are not confined to providing mystical pseudo-explanations; even a cursory look at the field of developmental biology today provides striking evidence that the quest to understand development beyond the genome is progressing apace. In other words, Rosenberg's preferred vision of developmental biology is not the only one – let alone the best one – available.

Rosenberg implausibly contends that a full explanation of development will have 'no room' for any reference to cell physiology, or anything else above the level of 'the molecular processes that subserve development'. He argues that, 'just as cell–cell signaling is ultimately to be cashed in for a chain of molecular interactions that extend from one stretch of nucleic acids to another across several lipid bi-layers (the cell membranes), all other cellular structures implicated in the machinery of differentiation will eventually have to be disaggregated into their molecular constituents, if development is fully to be explained' (Rosenberg 1997: 455, 454). However, it is not clear that such disaggregation constitutes an adequate explanation at all, though Rosenberg assumes that it does, for a microreduction may be no more explanatory than a macroreduction, especially if we do not adequately understand the mesolevel phenomena.

We cannot assume, as Rosenberg would have us do, that the background factors are computable as imagined. As this assumption is a hedging tactic to avoid triviality, we need not grant Rosenberg's conclusions about physicalist antireductionism, the prospects for explaining development, or the wondrous powers of the homeobox genes. (I return to the homeobox genes in Chapter 2.)

BEYOND THE HEDGE

The difficulty with the HHH in the context of development is that it amounts to paying lip service to development rather than taking it seriously. But what would it mean to take development seriously? I suggest that what we need is a better, less suspect variant of a context simplification heuristic, a more honest one, one more adequate to investigating biological reality, and one less likely to yield inference to an inappropriate conclusion about development. Following J.H. Woodger (Woodger 1952), I refer to this alternative interpretation of the second premise as the 'constant factor principle' heuristic (or CFPH).

Writing a half-century ago, Woodger noted the importance of heuristics in biological experimentation. For Woodger, as for others, the assumption of constant factors is often a useful simplifying strategy in order to achieve experimental tractability. In attempting to understand how genes function, for example, we may assume that the environment is a constant factor; against a constant environmental background, we may then solve for phenotypic differences by exploring the genotype, that is, the variable factor (Woodger 1952: 186). Where such differences are found, we may account genetically for the existence of variations. The heuristic assumption of constant factors is methodologically commonplace, but it is by no means infallible, as should be evident from the discussion thus far. Nonetheless, I will urge here that Woodger's 'constant factor principle', interpreted as an heuristic, works against the particular biases of the HHH and so is a more legitimate simplification heuristic and a more appropriate interpretation of our second premise.

Considering Woodger's own example permits a further bias of context simplification through holding factors constant to emerge. The strategy of solving for genes by holding the environment constant presumes that there are only two sources of variation: genetic or environmental. However, other potential sources of variation are stochasticity and epigenetic interactions, neither of which is, strictly speaking, genetic *or* environmental – they result from development as such. Especially instructive is the work of Gaertner, who, over a period of thirty years, developed *genetically identical* strains of laboratory mice and rats and reared them under *identical environmental* conditions – and yet the mice and rats were, nonetheless, phenotypically *non-identical*, thereby demonstrating the existence of a source of ontogenetic variation that was neither genetic nor environmental (Gaertner 1990; Molenaar et al. 1993). Thus, phenotypic differences against a constant environmental background may not legitimately be presumed to be genetically based (or environmentally based), even though some versions of context simplification heuristics simply do not guide us to investigate alternative possibilities.

But the most encompassing problem with simplification heuristics, especially as instantiated in hedgeless hedging, is the tendency to downplay or simply neglect the causal significance of those factors held constant. Consider loss-of-function experiments. A typical loss-of-function experiment is one in which, against a constant background, a particular gene is manipulated so that it is not expressed at the right time and place; the investigators then observe the phenotypic outcomes and conclude that the outcomes are caused by the misexpressed gene. However, often investigators will, in the absence of a complementary gain-of-function experiment, draw an additional, unwarranted conclusion, namely that the gene, when properly expressed, is itself causally responsible for the correct phenotypic outcome. This latter inference simply does not follow. As Keller notes, 'such an inference appears to make sense only to the extent that the entire physical-chemical apparatus of the organism and its environment are effaced' (Keller 1994: 90).[4]

Holding factors constant is a good and necessary part of proper science. *But effacing their causal importance is not.* It is for this reason that we should prefer the CFPH over other simplifying strategies as a methodological heuristic in making and interpreting experimental assumptions.

The CFPH asserts that, 'if, in a series of experiments, certain factors are constant, not necessarily in the sense of unchanging in time, but in the sense of being of the same kind in each experiment, then *nothing can be asserted on the basis of those experiments about the role of such constant factors in the production of the observed result*' (Woodger 1952: 186; italics added). Prohibited assertions, according to the CFPH, include claims that the constant factors '"play no part" in the processes involved', or that they play only a supportive role. Different experiments, perhaps even different sorts of experiments, are required for establishing the latter results; they cannot be inferred from scenarios in which the constant factors are never varied.

Immediately, then, we see that the usual conclusion (that is, 6) cannot be validly inferred if premise 2 is interpreted according to the CFPH. As long as premise 2 is interpreted as an invitation to hedge hedgelessly, then our near-universal presumption that genes are more causally relevant than other factors in development generates the conclusion that development is best explained as a matter of genes operating against a constant background of supportive conditions. However, if premise 2 is interpreted along the lines of the CFPH, then we are free to imagine (and explore) other scenarios for premise 5 and are thus less likely to imagine the validity of inferring the orthodox conclusion.

The second premise, now more satisfactorily interpreted according to the CFPH, reads as follows:

2'. Simplifying the context of a system (the definition of which is admittedly contingent) is advantageous if we want to learn about intrasystemic causal factors, but we must not neglect the possible importance of those contextual factors we abstract away.

Accordingly, we are invited to infer the following from premises 1 through 5, replacing 2 with 2':

6'. Therefore, against standard background conditions, aspects of organismal development may be partially a matter of gene action and activation, and it remains to be determined whether (and how) extragenetic factors make a specific causal contribution to ontogenesis.

Because of the limitations of the sorts of experiments undertaken thus far, we just do not know enough about development to conclude that the specificity of development is a matter of gene action and activation; and given the analysis to follow in later chapters, we will often have good reason to be suspicious of any such claim. An appropriate interpretation of premise 2, coupled with appropriate variations on the fifth premise, demands further, broader exploration of causal factors in development.

The CFPH is more satisfactory methodologically than either context simplification *simpliciter* or hedgeless hedging just because it provides grounds to avoid the biases of context simplification, and moreover because it guards against the particular biases of hedgeless hedging. However, though the CFPH is a better heuristic, it is itself subject to systematic bias. Woodger himself remarks that it has 'more than once been forgotten in connexion with genetical problems' (Woodger 1952: 186). Nevertheless, in cautioning against interpretive folly even while promoting the necessity of simplification, the CFPH is a superior guiding principle.

HEURISTIC SUPERIORITY

How does the CFPH work in practice? What is its 'cash value'? Michael Ruse once claimed that 'there is little if anything of value in Woodger's work, and that therefore the time has now come to draw a decent veil over a biological dead-end' (Ruse 1975: 2). Ruse was mainly concerned with overthrowing Woodger's peculiar axiomatisation of biological theory, though he commented on Woodger's discussion of biological methods as well. Nils Roll-Hansen extended this judgment of Woodger in a more extended analysis of the latter's methodological proposals. For Roll-Hansen, the nature of the biological dead end is that, by Woodger's lights, we are 'forced to consider organic wholes and their properties as the unanalyzable elements of biology' (Roll-Hansen 1984: 423). Were that in fact the case, the suggestion here that Woodger's constant factor principle is methodologically important would be woefully misguided. Allow me to show, then, rather than just state, that the CFPH is methodologically productive and important rather than a biological swan song.

Instructive in this regard is Ruse's peculiar discussion of simplifying assumptions, especially the heuristic importance of environmental control. Woodger notes that in the study of Mendelian heredity, 'only one class of environments is involved and is usually not even mentioned. Some interesting discoveries may await the investigation of multi-environmental systems'

(Woodger 1959: 427). Ruse's commentary is worth quoting at length:

> Such interesting discoveries may indeed lie around the corner; but it is hardly the case that conventional geneticists are ignorant of the effects of the environment (on development). Consider the following discussion which follows the introduction of the concepts of phenotype and genotype in a recent elementary textbook.
>
> > It is important to realise that an adult animal is the result of the interaction during development of the genes and the environment. If Mendel's tall pea plants had been grown under poor conditions while the short ones had been grown in the very best environment, the phenotypic appearance of the two could have been very similar. In conducting experiments on heredity it is therefore of paramount importance that, when comparing two or more types, they should be reared under identical conditions.[5]

Ruse resumes his commentary:

> Nor is it the case that there is something perverse in the admitted fact that geneticists, particularly population geneticists, tend to ignore the environment in their calculations. The problems of genetics are so complex that, so far, they have just had to make simplifying assumptions, particularly about the environment. But what scientist does not make simplifications? (Ruse 1975: 8–9)

It is true enough, as I have maintained from the outset, that scientists must make simplifying assumptions to get research programmes off the ground. However, what is at issue here is not the making of such assumptions; what is at issue is *which simplifying assumptions are made*, and *what inferences we are entitled to draw* given that simplifying assumptions are operative. To these two problems, Ruse makes no contribution. In fact, his discussion functions as an implicit justification for hedgeless hedging.

Ruse makes a further claim, namely that Woodger himself contributes nothing to the task of bringing environmental considerations into the practice of genetics (Ruse 1975: 9). To the contrary, given the necessity of holding some factors constant, Woodger's CFPH makes a significant contribution to this project. If nothing can be inferred about the causal contribution of those factors held constant in a particular experiment, then we are compelled to undertake different sorts of experiments, varying other factors serially and then integrating the results of the serial experiments.

It must be underscored that in conducting such serial experiments, we must be wary of the kind of simplificatory asymmetry Wimsatt cautions against in the use of particular heuristics. For as long as the factors to be varied are restricted solely to the class of systemic or intrasystemic variables (against a constant environmental or extrasystemic background), a systematic bias in favour of the model system's independence of the environment may emerge – and yet go unnoticed (Wimsatt 1980: 233; Wimsatt 1986c: 302–303). Thus a full application of the CFPH requires appreciating the insight that 'what one must control is a function of what relationships one is studying' (Wimsatt 1986c: 303) and also what we count as comprising our particular system.

If one is interested only in causal relationships independent of environmental context, then one conducts experiments in which the environment is held constant – which is fine, as far as it goes, although the CFPH cautions that interpretation of the results must be constrained by admission of the limits of the experiment. Such constrained interpretations are few and far between, though, as evinced in recent discussions of what we can expect now that the human genome has been sequenced. However, if one is interested in more complete causal analysis, the kind of analysis affording fewer and less onerous interpretive constraints – the kind of analysis legitimately yielding interpretations of real-world significance – then environments cannot be universally held constant.

There are several ways of proceeding toward this end through the CFPH, and I will briefly mention two. One is to adopt the perspective of methodological systemism (Mahner and Bunge 1997; Robert 2000c). Methodological systemism is a form of modest reductionism whereby we should reduce where possible, but never greedily (that is, reduce without sacrificing explanatory power), and we should expect – and account for – the material emergence of properties neither explicable nor predictable on the basis of lower-level properties.[6] Systemism functions as a sort of middle ground between reductionism and holism, according to which systems are ontologically and epistemologically irreducible to their *composition* (the component parts of a system), their *structure* (exogenous and endogenous), or their *environment* (immediate or proximate – those things related to but not part of a system's components). No biological system reduces to just one or two of composition, structure, or environment; rather it is emergent from all of them together.[7] Following Riedl, I interpret systemism as representative of complex causal structures, causes as related in positive and negative feedback networks. Riedl observes that 'if it is true that feedback cycles can connect levels of different complexity, such as the phenotype and the genotype, then we must accept a flow of cause and effect in two directions, up and down the pyramid of complexity. Then we should also accept causality in living beings as a system in which effects may influence their own causes' (Riedl 1977:

366). As a method, systemism requires theoretical and experimental attention to all aspects of the system under study – not all at once, of course, *contra* Roll-Hansen; serial experiments will do, as long as the variables cut across environmental, compositional, and structural levels.

Though examples of this sort of approach are relatively rare, one particularly nice instance is Daniel Lehrman's experiments on reproduction in ring doves (Lehrman 1965; Gray 2001: 201). The factors that Lehrman and colleagues serially varied were not restricted to any single system level: nest-building behaviour, 'doses' of male courtship, hormone levels, and so on. Having varied one factor, the investigators measured the effects on the system; then having varied another factor, they measured again. The end result is a more complete analysis of causation than would have been afforded by simply holding the environment constant. With more sophisticated statistical and experimental techniques, it is of course possible to vary more than one factor at a time, generating a still richer perspective.

A second method for generating similarly robust results is to engage in multidisciplinary investigation of the sort becoming commonplace in the new field of evolutionary developmental biology – the subject of Chapters 6 and 7. For instance, Brakefield and colleagues, in their work on eyespot patterns on butterfly wings, brought together tools from population genetics, evolutionary biology, ecology, developmental biology, and developmental genetics in a series of experiments that, taken together, provide an amazingly rich overview of the developmental mechanisms and evolutionary trajectories of this particular aspect of butterfly wing morphology (Brakefield et al. 1996; Brakefield and Kesbeke 1997; Brakefield 1998, 2001; Brakefield et al. 1998; Brakefield and French 1999; van Oosterhout and Brakefield 1999; Roskam and Brakefield 1999). Such an integrative approach is indicative of future prospects in understanding development – despite occasional ill-informed protestations that methodological reductionism is the only way forward.

Multileveled, multidisciplinary analysis – appropriately heuristically informed – is the surest route for generating results adequate to the complexity of the biological world, though from a comparatively simple, tractable, starting point. Such results are now beginning to be seen, and they will eventually enable us to have a fuller understanding not only of the roles of genes in development but also of organismal development as such – or so I will argue.

Even in applying a well-chosen heuristic to a particular problem, a crucial caveat to bear in mind is that the application of the heuristic may *transform* the problem into one for which an answer is available. Yet, as the new problem is 'nonequivalent but intuitively related' to the original problem, *we are no longer in fact solving for the original problem* (Wimsatt 1986c: 295). When the transformation goes unnoticed, we may believe we have indeed solved the original problem. We have not.

It is for this reason that the core problem of embryology is not, *pace* Rosenberg, entirely solved by modern developmental genetics. The translation of embryology's hard problem (how a specific complex organism arises from a single, relatively homogeneous cell) into a problem about gene action and activation generates explanations at the level of genes; but these explanations solve (or, rather, begin to solve) the subsidiary problem of the role of genes in development, not the problem of development as such (Robert 2001a). The trick is to integrate these explanations with other developmental (cellular, environmental, and ecological) explanations within a larger organismal framework, rather than to assume that we understand development because we are beginning to grasp gene function.

To take development seriously is to take development as our primary *explanandum*, to resist the substitution of genetic metaphors for developmental mechanisms; though some, perhaps many, developmental mechanisms will indeed be genetic mechanisms, others will be irreducible to genetic substrates. It may well turn out, even if we take development as our *explanandum*, that genetics will be our *explanans*; but we should not assume this a priori, and neither should we blindly aim for this result. There is indeed good reason to believe that genetics reduces to development, and not the other way around – but let's not get ahead of ourselves.

2

Exemplars

The amazing thing about . . . development is not that it sometimes goes
wrong, but that it ever succeeds.

– Veronica van Heyningen (2000)

Textbooks of developmental biology are rife with examples of what makes
development such a fascinating science. In this chapter, I discuss but three.
We will return to these examples in later chapters as we explore the nature
of development and its evolutionary significance, but their introduction here
gives us some signposts of significant events and achievements in the study
of development over the past 125 years.

The exemplars I have chosen are not necessarily the most groundbreaking
achievements of developmental biology – I am not quite sure how one would
select the most important ones. As will become evident, though, they are both
important and heuristically and rhetorically useful, and they well represent the
three elements of the title of this book: embryology, epigenesis, and evolution.

EMBRYOLOGY: ROUX AND DRIESCH

To set the stage for Chapter 3, and so to illustrate the contrast between preformation and epigenesis, it is useful to review several important experiments
undertaken in the early years of experimental embryology in the nineteenth
century. The experiments in question are those of Wilhelm Roux (1850–1924)
and Hans Driesch (1867–1941).

Roux's experiments, some of the first conducted on an embryo, were indeed
pathbreaking, and they are also amongst the most well-known experiments
in embryology. Roux was founder of the *Entwicklungsmechanik* (developmental mechanics) program – the first physiological approach to the study of

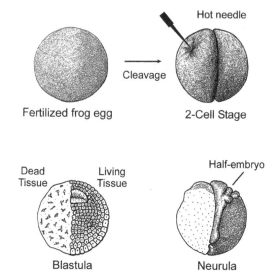

Figure 1. Roux's half-embryo experiments. Top left: A fertilised frog egg prior to cleavage. Top right: After cleavage at the two-cell stage, the left blastomere is pricked with a hot needle. Bottom left: The blastula consequently contains both dead (left) and living (right) tissue, the dead tissue not having been separated. Bottom right: The blastula develops into a half-embryo at the neurula stage. Redrawn with permission and modified from Figure 3.14 in Gilbert (2000a: 58).

embryology (Roux 1894) – though he was not single-handedly responsible for the emergence of *Entwicklungsmechanik* (Maienschein 1991b).[1]

Roux held to a version of preformationism, and he propounded a view of development known as mosaic development, according to which nuclear materials hive off qualitatively into different daughter cells during cell division: each resulting piece of the organism contains a different bit of nuclear material, though each individual, semi-independent piece is also an integral part of the whole (of the larger picture, as it were). In 1888, Roux attempted to test his hypothesis of mosaic development.

Roux hypothesised that an embryo at the two-cell stage will have the determinants of the left side of the organism in one blastomere, and those of the right side in the other blastomere. Thus, if one were to kill one of the two cells at this stage, the embryo would retain only half of the determinants of the organism and should develop into only a half-embryo. Roux therefore killed one of the cells of a two-celled frog embryo by using a hot needle (see Figure 1). The other cell continued the usual cleavage process apparently independently of the dead cell, and a half-blastula resulted, which then experienced an abnormal gastrulation, producing in the end what Roux interpreted as a half-embryo. Hence Roux believed he had experimentally vindicated his hypothesis that the embryo is a mosaic of cells, each able to produce only a specific part of the developed organism. Roux continued to experiment on embryos and became convinced of mosaic development; when he achieved results that contradicted his hypothesis, as Maienschein (1991b: 51) reports, he elaborated adjunct hypotheses to protect his core belief.[2]

Hans Driesch expected to confirm Roux's results in a series of experiments performed several years later. Driesch was working in Naples at the zoological station, where sea urchins were abundant, so he conducted his studies with sea urchin embryos.

Roux's methodology in 1888 may have been flawed in that he did not remove the dead cell; the behaviour of the other cell may have been influenced by the presence of the dead cell, rather than manifesting mosaic development (Maienschein 1991b: 50). It is noteworthy, therefore, that Roux had been unable to separate the blastomeres, though Oscar and Richard Hertwig had shown in 1887 that vigorous shaking in water would suffice.[3] Thus Driesch separated his sea urchin blastomeres in an effort to confirm Roux's results on mosaic development (see Figure 2).

The next day Driesch found, to his surprise, that the separate blastomeres had each developed into 'typical, actively swimming blastulae of half size'.[4] That is, they had become, not half-embryos, but half-sized embryos. The blastomeres therefore remained totipotent, able to respond to (intraorganismal) environmental conditions and to transform themselves accordingly. Each cell retained the ability to regenerate whatever material went missing in the separation-by-shaking of the two blastomeres (Maienschein 1991b: 51, 52; also see Gilbert 2000a: 59–61). Rather than becoming a differentiated future part of the organism, each blastomere is able to regulate its development in order to produce a whole (not a half) organism. Hence the epithet 'regulative development', in contrast to Roux's mosaic development.

Driesch allowed for both mosaic and regulative development and did not initially emphasise the differences between his results and those of Roux. Had Roux been able to separate the frog blastomeres, rather than permitting the dead one (which Driesch thought may in fact not have been dead after all, but rather merely maimed) to remain in contact with the live one, perhaps the embryo would have developed normally after all, as was the case with Driesch's sea urchins (Maienschein 1991b: 52). Later, though, Driesch would contend that there was a vast difference between his and Roux's results, at which point he renounced the study of embryology and set out to produce an antipredeterminist, antimosaic, vitalistic philosophy of development.

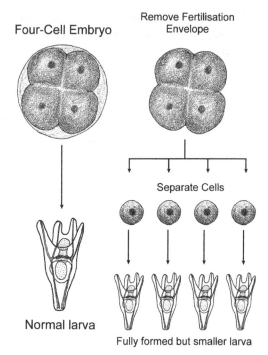

Figure 2. Driesch's miniature embryos. Top left: A four-celled sea urchin embryo. Bottom left: The embryo develops into a normal-sized pluteus larva. Top right: A four-celled sea urchin embryo removed from its fertilisation envelope. Middle right: The embryo is shaken and so separated into its four constituent cells. Bottom right: The resulting pluteus larvae are smaller than normal, but otherwise fully formed (though non-identical). The five larvae are drawn to the same scale. Redrawn with permission and modified from Figure 3.15 in Gilbert (2000a: 59).

These experiments provide a sense of the different theories of development in circulation before the rediscovery of Mendel's papers and the onset of classical genetics: Roux represents a kind of preformationism whereas Driesch represents a kind of epigenesis, the themes of Chapters 3 and 4.

EPIGENESIS: THE HOMEOBOX GENES

Epigenesis means, simply, development. However, epigenesis is usually understood as the antithesis of preformationism, referring to a pseudo-mystical doctrine of the emergence of complexity as the result of some unidentified guiding force. As indicated in Chapter 4, Waddington attempted to achieve a non-mystical account of development by melding epigenesis with genetics,

thereby coining 'epigenetics' as a new staple in the developmentalist's lex-
icon. Epigenetics in its most recent usage refers primarily to the regulation
of gene expression. Gene regulation (epigenetics) is clearly important in understanding development (epigenesis), but as I explained in Chapter 1, gene
regulation is not all there is to understanding or explaining development, so
epigenetics and epigenesis are not identical. Despite occasional missteps to
the contrary (e.g., Rosenberg 1997), a good place to start in understanding
the regulation of gene activity is with the homeobox genes (Robert 2001a).

The homeobox genes are a highly conserved class of genes involved in the
regulation of cell pattern and the regulation of genes involved in the establishment of basic body plans in animals and plants. The homeobox is a sequence
of 183 nucleotides encoding 61 amino acids. The amino-acid-specified home-
odomain is a DNA-binding domain regulating specific DNA–protein interactions, which influence DNA transcription. The homeobox, shorthand for
'homeotic box', builds on William Bateson's (1894) notion of homeosis, according to which part of an embryo is transformed (in development) into
another structure. Foundational work on homeobox genes was conducted by
Edward Lewis (1978; also see 1994) and Walter Gehring (McGinnis et al.
1984; Gehring 1985; for a popular account, also see Gehring 1998), among
many others.

Experimental manipulations of homeobox genes generate amazing – even
grotesque or monstrous (Rehmann-Sutter, 1996) – results. Consider homeotic
mutations on the third chromosome in *Drosophila*. Within the Antennapedia
complex, a mutation in *Antennapedia* converts antennae into legs; within the
Bithorax complex, changes in *Ultrabithorax* expression, such as deletion or
mutation of the gene or its regulators, effectively transform the third thoracic
segment into another second thoracic segment, and so produce the replace-
ment of halteres (balancers) with a second set of wings (Figure 3).

Homeobox genes are often referred to as master control genes, setting in
motion a complex of processes necessary for the formation of, for example,
heads or limbs (see, for instance, Gehring 1998). However, as I have argued
elsewhere (Robert 2001a), and as I reiterate in what follows, homeobox genes
are better construed as efficient micromanagers in development rather than
as master controllers. This is because homeobox genes are no less regulated
than other genes in development, whether by cell–cell signalling, hormones,
or other means. Moreover, whether homeobox genes can be manipulated
to produce large-scale changes in development – such as the construction
of ectopic limbs or the rearrangement of body plans – is crucially context
dependent. As Akam has noted, a misexpressed homeobox gene may well lead
to a new pattern ectopically, but only if the appropriate downstream targets are

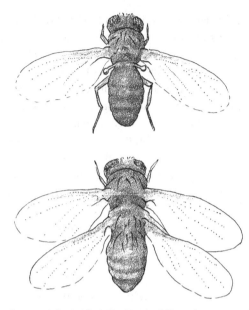

Figure 3. Homeotic mutant in the fruit fly, *Drosophila melanogaster*. Top: A normal fly with a pair of halteres on the third thoracic segment and a pair of wings on the second thoracic segment. Bottom: An *Ultrabithorax* mutant with two pairs of wings instead of the normal single pair and pair of halteres, as the third thoracic segment is transformed into another second thoracic segment. Redrawn and substantially modified from the photographs in Figure 2.3 in Carroll et al. (2001: 21).

present at the new site: 'When it comes to the downstream targets of the *Hox* genes, context is everything, in particular, which other transcription factors are present in the same cell will be a key factor determining the outcome of *Hox* gene action' (Akam 1998: R678).

The master control gene trope significantly oversells the role of homeobox genes in development; it also tends toward overestimation of the capacities of homeobox genes in effecting evolutionary change. (I discuss one example in some detail in Chapter 7; also see Robert 2001a.) Graham Budd (1999: 327, 329–330) has proposed a plausible alternative model of homeobox activity according to which evolutionary change is not initiated or driven by changes in (the timing of) the expression of these genes; rather, given that homeobox genes offer an efficient way to channel developmental information and build particular body plans, Budd proposes that homeobox genes are used post hoc to streamline developmental processes once gradual morphological change has occurred. He calls this model 'homeotic takeover' (see Figure 4).

It is well established that the same phenotype can be generated with a number of different phenotypes under the same or variable environmental

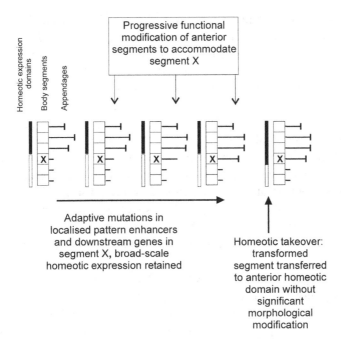

Figure 4. Homeotic takeover. Depiction of the gradual addition of a new trunk segment (segment X) in arthropods by means of morphological anticipation of homeotic transformation. Adaptive changes in regulators and downstream targets of homeotic genes alongside progressive functional modifications in segments anterior to X allow for the new segment to be taken over by anterior homeotic domain without any significant functional or morphological disruption. Redrawn with permission and modified from Figure 2 in Budd (1999: 330).

conditions. Accordingly, Budd notes that homeotic takeover is related to Waddington's notion of genetic assimilation, whereby, over time, nongenetically (e.g., environmentally) induced aspects of morphogenesis may be assimilated into the genome (Budd 1999: 329, 331; Waddington 1961). Waddington's genetic assimilation experiments with *Bithorax* indicate that flies with two pairs of wings (instead of a set of wings and a set of halteres) can be produced not only by loss of function mutations in *Ultrabithorax* but also by polygenic changes spread across multiple chromosomes. These polygenic changes can be imagined to occur and accumulate over a number of generations, each change involving minor mutations or the revelation of hidden genetic variation, and then being fixed in the population either through drift or selection (or both). The development of *bithorax* phenocopies so produced could then be 'taken over' at the *Ultrabithorax* locus without dramatic morphological or functional change (Budd 1999: 329).

An important difference between genetic assimilation and homeotic takeover is that the former is geared toward progressive, gradual genomic control, whereas the latter involves a quantitative shift from downstream to more upstream developmental genes (Budd 1999: 331). I have suggested elsewhere (Robert 2001a) that homeotic takeover is a token of Weiss and Fullerton's (2000) more general notion of 'phenogenetic drift' (which is also clearly related, but not identical, to genetic assimilation). Phenogenetic drift refers to the process of different genotypes being associated, over time, with the same phenotype, through genetic substitution. Because selection acts on phenotypes (not genotypes), these different genotypes will be evolutionarily indistinguishable (that is, indistinguishable to natural selection). Weiss and Fullerton propose that, as against Dawkins-style Neo-Darwinism, 'genes may be better understood as having a phenotypic *raison d'être*' (2000: 188, 189). If this is true of genes generically, then it is true as well of homeobox genes specifically.

In this regard, consider Newman and Müller's (2000) idiosyncratic but ultimately very useful interpretation of the relationship between genetics and epigenesis. Newman and Müller invoke a 'pre-Mendelian' (or pre-genetic) interpretation of 'epigenetic mechanisms' whereby these are 'conditional, non-programmed determinants of individual development' such as tissue–environment (both exogenous and endogenous) and tissue–tissue interactions (Newman and Müller 2000: 305–306). For Newman and Müller, 'as evolution proceeds, genetic change that favors maintenance of morphological phenotype in the face of environmental or metabolic variability co-opts the morphological outcomes of epigenetic processes, resulting in the heritable association of particular forms with particular genealogical lineages'. On this alternative interpretation of epigenetics (alternative, that is, to the modern usage specifying gene regulation exclusively), 'the correlation of an organism's form with its genotype, rather than being a defining condition of morphological evolution, is a highly derived property' (pp. 306, 304).

As morphological change proceeds gradually and epigenetically (in Newman and Müller's idiosyncratic sense), homeotic takeover, as an instance of the more general phenomenon of phenogenetic drift, provides a model for homeobox genes as efficient micromanagers advantaged by natural selection.

We don't need the baggage of 'master control genes' to understand or explain the established (and the hypothetical) roles of homeobox genes in development and evolution. We can manage just fine with less ideologically inflated notions. Accordingly, we should not allow ourselves to be swept up in the hoopla of 'hoxology' (the term is borrowed from Gould 2002:

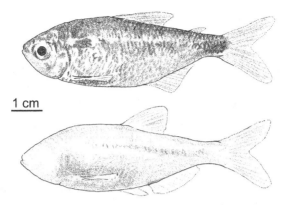

Figure 5. *Astyanax mexicanus*, the Mexican tetra fish. Top: Pigmented surface fish with eyes. Bottom: Unpigmented blind cave fish. Note the other minor differences in morphology (especially the fins) between the two fish. Redrawn and significantly modified from the photographs in Figure 1 in Jeffery (2001: 4).

82 *et passim*). The lessons of Chapters 3 and 4 are similar: we don't need genetic programmes or instructions, or even specifically genetic information, in order to understand and explain the developmental effects or evolutionary significance of genes.

EVOLUTION: BLIND CAVE FISH

The final chapters of this book explore interrelations between development and evolution and between developmental and evolutionary explanations. Along with many others (Raff 1996; Arthur 1997; Hall 1999; Wagner 2000), I maintain that understanding development is important in understanding evolution, just as understanding evolution is important in understanding development. Consider, then, a case involving evolutionary alterations in development that lead to marked heritable phenotypic change in the Mexican tetra fish, *Astyanax mexicanus* (Yamamoto and Jeffery 2000; Pennisi 2000; Jeffery 2001).[5]

The fish exist in a few dozen isolated populations in northeastern Mexico; some live in streams and ponds, whereas others live in caves and underground pools. Over the past million years or so, and amongst other evolutionary changes, the cave fish have gone blind whereas those who live above ground continue to have large eyes. The cave fish have also lost pigmentation whereas the surface fish have not (Figure 5).

As Yamamoto and Jeffery (2000) have observed, although eye development starts off in the usual way in the cave dwellers, producing a rudimentary lens and optic cup, after twenty-four hours the cells in the embryonic lens die, the cornea and the iris fail to develop, and the retina does not organise into distinct layers. Eventually, the eyeball sinks back into the skull and is covered by a skin flap. Yamamoto and Jeffery (2000) were able to show that the lens is responsible for promoting eye development, as a cave fish embryo which, at twenty-fours hours of age, has had the lens vesicle of a surface-dwelling conspecific transplanted into its optic cup develops a large eye with a distinct pupil and a properly developing retina; its other, untreated eye, sinks into its orbit. Thus, 'a surface fish lens can induce the development of anterior eye parts that have been lost during cave fish evolution' (p. 631).

So, loss of eyes in *Astyanax* occurs by disruption of the developmental pathway at a specific point, and it is possible to rescue eyes by providing a signal at the right time and place, for no developmental potential has been lost. Invoking the developmental phenomena, in addition to ecological and population genetic aspects, thus helps to complete the explanation of the evolution of these two tetra fish morphs and so illustrates one way in which development and evolution interrelate.

Jeffery has recently argued that these cave fish are an excellent model system for evo–devo, not least because, as cave fish descend from their surface-dwelling conspecifics, we already know the direction of developmental change, which makes for an easier reconstruction of the evolutionary history of the cave fish (Jeffery 2001). Moreover, studies of cave fish in relation to surface fish highlight 'the possible role of tradeoffs between constructive and regressive processes' in evolution and development (p. 9). However, much work remains to be done. For instance, it has been suggested that the cave fish have 'traded in' their eyes – which are not needed in the underground streams, and so their retention is not subject to evolutionary pressures – for other, more adaptive features, as the troglodytic fish have more teeth and taste buds than their surface-dwelling conspecifics (Vogel 2000). It is possible (even likely) that the surface-dwelling, large-eyed fish are developmentally precluded from evolving improved gustatory and masticatory apparatuses; such a result, if it can be established, would be important in showing how development biases evolutionary outcomes. However, if the surface-dwelling morph's variability is not biased against or away from this innovation, we can still see that the developmental and evolutionary co-production of cave fish blindness remains evolutionarily significant and demands joint developmental and evolutionary explanation.

CONCLUSION

There are many examples of fine achievements in developmental science. We might have dwelled on those of E.G. Conklin, Hans Spemann, Viktor Hamburger, Christianne Nusslein-Volhard and Eric Weishaus, Gilbert Gottlieb, or Paul Brakefield. However, my aim in this chapter was not to reconstruct the many momentous events in developmental biology since the nineteenth century; my aim instead was to introduce the major themes of this book in an illustrative rather than comprehensive way. These are as follows: that the ancient dispute, from Aristotle onward, between preformationism and epigenesis, evident in the examples of Roux and Driesch, is still with us and demands attention; that tall tales about the significance of genes may be brought down to size by theoretical articulation and experimental execution; and that those who would insist that development and evolution have nothing to say to each other (e.g., Wallace 1986) are just plain wrong. Let us now attend in a more programmatic way to the imperative to take development seriously.

3

Scylla and Charybdis

> The preformation idea has always led to immediate, if temporary suc-
> cesses; while the epigenetic conception, although laborious, and un-
> certain, has, I believe, one great advantage, it keeps open the door for
> further examination and re-examination. Scientific advance has most
> often taken place in this way.
>
> – *Thomas Hunt Morgan* (1909)

A central contention of this book is that our understanding of biology, and the
very nature and history of living things, hinges crucially on our understanding
of development as the basic biological process. Development is what distinguishes biological systems from other sorts of systems, and it is the material
source of evolutionary change.

In unpacking my claims about development, it is useful to explore a number
of metaphors used in attempts to explain development: these include 'information', 'programme', and 'triggering'. Rather than addressing these in turn,
I instead offer an account of how these metaphors, and others, come together
in what I call the 'modern consensus' on development.

Our very sense of biological possibilities, and of the nature of gene action and activation, is constrained by our conceptions, whether implicit or
explicit, of epigenesis and preformation. The idea of epigenesis has a rich
history dating back to Aristotle, and it is typically understood as the antipode
of 'preformation'; thus, even though there are no pure preformationists or epigenecists still with us, both epigenesis and preformation will initially occupy
me in this chapter.[1] That discussion will serve as the basis for my characterisation of the modern consensus on development and for efforts in later
chapters to diagnose its failings.

In Greek mythology, Scylla and Charybdis lived opposite one another in
the Strait of Messina. Whirlpool-like Charybdis would swallow the sea and

spit it back out, while monstrous Scylla would attack passing sailors with her six heads (each with three rows of teeth!). Sailing safely between them was no mean feat, as Odysseus' companions learned all too well, and the phrase 'Scylla and Charybdis' has since come to refer to a pair of equally bleak options. Thomas Hunt Morgan (1907: 384) once described epigenesis and preformation as Scylla and Charybdis; throughout the twentieth century, commentators took on the pseudo-Odyssean task of attempting to fall prey to neither. But, unlike Odysseus, they did not seek a safe route between them; rather, they attempted to meld them.

Many have claimed success, such that development is now standardly construed as the epigenesis of something preformed in DNA. Accordingly, we are all presumably both epigenecists and preformationists. However, if, as so many have argued, neither epigenesis nor preformationism is correct, then, to my mind, a monstrous hybrid should be no better – combining Scylla with Charybdis merely multiplies our woes. We would be far better off navigating a harmless route through the Strait – or so I will argue. Our first task, then, is to identify the obstacles to safe passage. As there is a range of opinion as to exactly what 'epigenesis' and 'preformation' are supposed to mean, we must take care to understand them.

PREFORMATION VERSUS EPIGENESIS

The concept of preformation is supremely plastic, hurled as an epithet by one camp and updated for the modern world by another. It is generally agreed that *something preformed* develops (epigenetically) into a mature organism; as Løvtrup suggests, 'that something is "preformed" at the outset of each individual case of ontogenesis is so evident that it seems incredible that it has sometimes been thought necessary to supply experimental evidence in support of this point' (Løvtrup 1974: 8). Nevertheless, incredulity notwithstanding, *what* exactly that preformed something is, and *how* it is preformed, is a matter of dispute. It is worthwhile, then, to take a brief detour through the historically unhappy positions that have now presumably been neutralised through our understanding of the role of genes in development.

Both preformation and epigenesis, as concepts, date back to ancient Greece. In *De Generatione Animalium*, Aristotle, the first epigenecist, dismissed Hippocrates' preformationist idea that the bigger parts of an embryo appear earlier than the smaller parts not because they are formed earlier but because their size makes them easier to see. Aristotle was able to demonstrate that the heart in the chick embryo appears sooner than the lungs, and yet the lungs are bigger (and thus should be visible sooner than the heart); therefore as it is visible earlier in development, the heart must have been formed prior to the lungs, and not contemporaneously with them (Aristotle 1953: 734a).

Of course, Hippocrates' version of preformationism does not exhaust the category, for there are many ways of being a preformationist. One may insist with Hippocrates that the embryo as such is a tiny, perfectly formed adult needing but fire and food to grow into an adult; views popular in the seventeenth and eighteenth centuries go a step further, from the embryo to the sex cell: many naturalists held that a future organism is coiled in the sex cell, and then merely evolves into its mature form – one was therefore an animalculist (spermist) or ovist depending on one's sex-cell orientation. The eighteenth-century ovist Charles Bonnet refined the latter position by dispensing with 'minuscule men' in favour of the conviction that the 'germ' is a 'loose sum of all the "fundamental parts" of the future individual' (cited in Pinto-Correia 1997: 58). Still more recently, this sort of an account has been reissued in the garb of information theory, such that what is now considered to be preformed is genetic information. However, there are no pure preformationists any more, even in this attenuated sense.

But neither are there any pure epigenecists in the world today; that is, no one would seriously argue that a complex organism emerges magically (as it were) from a primitive homogeneous mass. Aristotle's basic insight about epigenesis – that the appearance over time of structures in the developing organism ought to be interpreted as evidence not merely of growth but rather also of change (development) – remains valid to this day, but it has nonetheless often been misinterpreted as if it required a vital driving force. (This is a misunderstanding of Aristotelian entelechy; Vinci and Robert, in preparation.) The basic insight remained stable and important for over two millennia, though in a variety of guises. Thus William Harvey's perspective on epigenesis, according to which the unformed (un*pre*formed) organismal substance takes up a form that is in it potentially, but not actually, has much in common with the various theses of Aristotle. So too does the more preformationistic perspective of Leeuwenhoek, whereby an organism takes up a (preformed) form that was there only potentially, not actually, requiring as it does a stimulus for its expression (Pinto-Correia 1997: 3, 85).

But Aristotle's notion of an entelechy, however misinterpreted, led to his guilt by association with those who posited vitalistic accounts of epigenesis (e.g., Müller 1996). To explain development from the relatively homogeneous, unstructured gametes through the increasingly complex organism, early epigenecists (without having recourse to the easy – though ultimately

36

mistaken – answer provided by their preformationist opponents) typically had to invoke some teleological force from without – some *vis essentialis* or *élan vital*, for instance. Epigenesis and vitalism have therefore been almost constant companions. Given the demise of vitalism, more recent epigenecists have sought form and structure elsewhere, usually within the organism itself. However, the best of the lot typically settle, sadly, on the nevertheless *animistic* (and otherwise problematic) idea of a genetic programme.

PREFORMATION AND EPIGENESIS: THE MODERN CONSENSUS

Let us presume, despite occasional slips in the literature to the contrary, that no one today holds that epigenesis is just the (actual and not merely visual) naturally unaided appearance of novelty in the unfolding of the 'information-ally preformed' organism. As is evident from even a cursory review of the literature on gene activation and regulation, a genome does not run the developmental show. However, such incisive thinkers as even Jacques Monod are prone to error on this count. Monod, himself discussing the debates between preformationists and epigenecists, argues that

> no preformed and complete structure preexisted anywhere; but the architectural plan for it was present in its very constituents. It can therefore come into being spontaneously and autonomously, without outside help and without the injection of additional information. The necessary information was present, but unexpressed, in the constituents. The epigenetic building of a structure is not a *creation*; it is a *revelation*. (Monod 1971: 7)[2]

In this passage, Monod suggests that developmental information, contained in toto in the genes, manifests by self-activation in organismal development.

It is not entirely apparent what Monod means here, though, for surely the co-author (with François Jacob) of the *lac* operon model of gene *activation* understands the necessity of 'outside help' in the actualisation of genomic potential. Yet the deep dispute is not over outside 'activation', which is granted, though sometimes not taken seriously enough, by everyone concerned; rather it is over the remainder of the conjunction, regarding the 'injection of additional information'. In other words, it is not in dispute that insofar as genomic potential is in fact actualised, the 'activation' – as it were – of the genome is context dependent, 'triggered' – as it were – by some extragenomic developmental component (the interactionist consensus).

In fact, the thesis of development-as-unfolding ('evolution') may be better construed as a relatively sophisticated kind of preformationism – one capable of integrating an explanation of the observable changes in an organism during ontogenesis – rather than as genuine epigenesis. For quite some time, only epigenecists could explain this observation, whereas preformationists were stuck with a theory about growth and no more. Now, however, preformationists could offer an equally plausible account by suggesting that the adult structures are not physically present (fully formed but very tiny) in the embryo; rather only the (preformed) genetic potential for those structures is present. It is this latter kind of preformationism that underwrites careless talk about genetic determinism, helping to substantiate the recent charge that gene centrists are merely preformationists in modern garb (Oyama 2000b; Mahner and Bunge 1997). Of course, there are no self-avowed preformationists among the ranks of geneticists, developmental biologists, and philosophers of biology. They rather see themselves as offering a preformationist–epigenecist hybrid, what I call the modern consensus.

As with the interactionist consensus, in which nature (genes) versus nurture (environment, experience) is superseded by the view that nature and nurture are *both* required to effect development, the modern consensus combines preformation and epigenesis rather than seeing them as in competition. Ernst Mayr is a leading exemplar of this view. He provides the following glossary entries of the notions of interest: 'preformation' refers to the theory 'that an embryo develops from material in which the essential form of the adult is "preformed", that is, already exists in its essential structures', whereas 'epigenesis' is the theory 'that new structures originate during ontogeny from undifferentiated material with the help of a vital force [*vis essentialis*]'. So construed, Mayr is correct to refer to both positions as 'now-discredited' (Mayr 1997: 310, 307). However, he also claims that the positions are 'partly right and partly wrong' (p. 156), requiring the advances of twentieth-century genetics and molecular biology to resolve the problem of development once and for all:

> The first step came from the field of genetics, which distinguished between a genotype (the genetic constitution of an individual) and a phenotype (the totality of the observable characteristics of an individual) and showed that during development the genotype, by containing the genes for becoming a chick, could control the production of a chick phenotype. By thus providing the information for development, the genotype is the preformed element. But by directing the epigenetic development of the seemingly formless mass of the egg, it also played the role of the *vis essentialis* of the epigenesis.
>
> Finally, molecular biology removed the last unknown by showing that the genetic DNA program of the zygote was this *vis essentialis*. The introduction

of the concept of a genetic program terminated the old controversy. The answer was thus, in a way, a synthesis of epigenesis and preformation. The process of development, the unfolding phenotype, is epigenetic. However, development is also preformationist because the zygote contains an inherited genetic program that largely determines the phenotype. (Mayr 1997: 157–158)[3]

This passage contains the central elements of the modern consensus on development.[4] These are the overlapping and mutually reinforcing theses of genetic informationism, genetic animism, and genetic primacy.

Genetic informationism is the position that genes contain the entirety of the preformed, species-specific developmental 'information'. *Genetic animism* refers to a genetic programme in the zygotic DNA controlling the development of an organism. *Genetic primacy* envisions that the gene is the unit of heredity, the ontogenetic prime mover, and the primary supplier and organiser of material resources for development, such that the phenotype is the secondary unfolding of what is largely determined by the genes.

These theses, taken together, comprise a modern, DNA-era reconciliation of preformation with epigenesis: the (preformed) genetic program in the zygotic DNA is transmitted between generations, contains all specific ontogenetic information, and determines the (epigenetic) development of an organism. There is no need for the previous centuries' preformationists' 'strange tales of small men' in reproduction (Pinto-Correia 1999), or for the epigenecists' vital force acting from without – the very ideas of genetic information and genetic programme solve this seminal problem of embryology.

There are, unfortunately, several difficulties with this set of views – which is one reason I prefer Newman and Müller's (2000) pre-genetic account of epigenetics discussed in Chapter 2. For instance, genes are not informational in the way supposed, nor do they initiate or direct ontogeny; there is no such thing as a genetic programme; and there is no straightforward 'unfolding' relation from genotype to phenotype. The modern consensus, melding epigenesis with preformation in light of modern genetics, is lacking on all counts – yet it persists in the writings and research programmes of a wide range of biologists. Before proceeding, in the following paragraphs and in the next chapter to offer arguments against the theses of the modern consensus, let me first justify characterising this cluster of views as in fact a consensus position.

The passage cited herein from Mayr is by no means idiosyncratic. For instance, in 1970 Fraser conjectured that 'the preformed basis of an individual has its identity in the constant informational content of the genes (the homunculus has a nucleic acid morphology). The epigenetic translation of the genetic information involves complex sets of genes acting in a variable

39

milieu of genetic and environmental effects, such that constant progression and fixed end points eventuate' (Fraser 1970: 57). Furthermore, Maienschein cites the position of the Medawars as 'fairly typical': 'the genetic instructions according to which development proceeds are indeed preformed, but their realization is *epigenetic*, i.e. turns upon influences acting upon the embryonic cell from the outside' (Maienschein 1986: 101–102, citing Medawar and Medawar 1977). As the Medawars said later in a slogan, 'genetics proposes, epigenetics disposes' (Medawar and Medawar 1983).

Similarly, J.A. Mazzeo, in his Introduction to the 1977 edition of Oscar Hertwig's classic 1896 book *The Biological Problem of To-Day: Preformation or Epigenesis?*, writes as follows:

> In our own time, the progress of molecular biology has finally elucidated that structure of the gene, a one-dimensional segment of DNA, which can both duplicate itself and serve as a 'template' for intermediary substances, messenger RNA and transfer RNA, which build the three-dimensional structure of the protein, and the organism is understood as the 'translation' of the information contained in the gene. The gene is, thus, a 'message,' and the truth of preformation is that what is 'preformed' is the information for making an organism. (Mazzeo 1977)[5]

So, in modern incarnations of preformationism, miniature encapsulated adults or their parts have been replaced by coded information or instructions contained within a genetic programme, executed epigenetically.

Two further instances of the modern consensus, construing the reconciliation of preformation and epigenesis as the epigenetic triggering of preformed genetic information, are worth noting. Gould remarks that

> The solution to great arguments is usually close to the golden mean, and this debate is no exception. Modern genetics is about as midway as it could be between the extreme formulations of the eighteenth century. The preformationists were right in asserting that some preexistence is the only refuge from mysticism. But they were mistaken in postulating preformed structure, for we have discovered coded instructions. (It is scarcely surprising that a world knowing nothing of the player piano – not to mention the computer program – should have neglected the storage of coded instructions.) The epigeneticists, on the other hand, were correct in insisting that the visual appearance of development is no mere illusion. (Gould 1977: 18–19)

Moreover, in a footnote in his translation of Aristotle's *De Generatione Animalium* – the *locus classicus* of epigenesis – A.L. Peck describes the triumph of epigenesis over preformationism as an overdue vindication of Aristotle. Aristotle, Peck maintains, was right to focus attention on the emergence of qualitative novelty in development. However, he notes that preformationism is not and was not entirely barren: 'like many erroneous theories, preformationism contained some truth, for we know to-day that the course of the embryo's development is predetermined by its genetic constitution' (Peck in Aristotle 1953: 145, note a).

But do we in fact know this? Is it really the case that 'the course of the embryo's development' is in some important way 'predetermined by its genetic constitution'? Surely that is the position of Mayr, and of Rosenberg, too, as I suggested in Chapter 1. It seems, as well, to be that of Monod, and the other commentators already cited. Even so, I remain sceptical.

UNPACKING THE MODERN CONSENSUS

There are two distinct perspectives on the idea of the external 'triggering' of genomic potential: one compatible and another (Monod's) incompatible with the ontogenetic requirement of specific environmental information – that is, information beyond the bare instruction 'switch on the preformed genetic program for development now' and its ilk. My sense is that most theorists hold to the former, not the latter, position, though in moments of weakness some theorists blur the distinction. However, I want to suggest in this chapter that even the former perspective is misguided in many of its numerous instantiations; all the while, I will contend that the latter perspective is surely mistaken as well.

A few distinctions, implicit in what I have said so far, should be made explicit. The view of epigenesis as *unaided evolution* (or unaided unfolding) is rather a sort of pseudo-epigenesis, one not currently represented in the biological or philosophical literature but one that might have characterised some earlier views. The euphemism of 'triggering' is used in two distinct ways. According to one, epigenesis is a matter of *initial triggering*, whereby the whole of the developmental potential resides in the genome, requiring mere activation from without for the production of an organism; according to the other, epigenesis is a matter of *contextual triggering*, a more sophisticated position according to which, although developmental potential resides exclusively in the genome, its actualisation occurs over time as the result of many external and internal activations or regulations (or both). The core idea underlying both of these positions is represented in the following passage:

> One can surmise that there must exist a machine to interpret the content of a DNA sequence. This machine is a preexisting living cell, made of a large

41

but finite number of individual components organized in a highly ordered way. Given such a machine, a fragment of DNA, provided it contains a well-formed sequence, is necessary and sufficient to produce a specific behavior representing part or all of the structure and dynamics of an organism. (Danchin 1996: 107–108)

Where epigenesis amounts to initial triggering, the cellular 'machine' triggers the production of the 'behaviour' pre-specified in the DNA fragment; where epigenesis amounts to contextual triggering, the cellular machine, in reactive interplay with the DNA fragment, regulates the expression of the preformed genetic information in the production of the specific behaviour. The latter position is, as I take it, the standard account of ontogenesis and can be made to fit all three theses of the modern consensus on development.

M. Moss characterises this sort of view as follows: 'at fertilization the diploid genome contains all the information necessary to regulate (or "cause") individual ontogenesis, requiring only an appropriately permissive and supportive environment for full genomic expression to occur' (Moss 1981: 366; Moss is, I should note, critical of such a perspective). The attentive reader will recognise this sort of view as substantially equivalent to the initial conclusion drawn in Chapter 1 about how best to understand and explain development, a conclusion that I maintain is mistaken.

Contrast this sort of modern consensus perspective with that of Schlichting and Pigliucci:

> It seems clear that, no matter how strong our belief in the power of reductionism as an explanatory scheme, the nature of the phenotype of any organism cannot be mechanistically deduced, even if we possess a complete DNA sequence of its genome. The elucidation of the utterly fascinating (and mind-numbing) gymnastics that comprise transcription and translation *has definitively crushed that hope*. Thus 'emergence' arises somewhere between DNA and the phenotype. This black box, often referred to as epigenetics, is now only being perforated with numerous small holes, shedding some dim light on portions of its contents. (Schlichting and Pigliucci 1998: 27; emphasis added)[6]

I will need to invoke yet another euphemism to account for (a) the idea that specific ontogenetic information is dispersed throughout the developing system and environment and therefore not localised (exclusively) in the genome; and (b) the further idea that 'activation' does not quite capture the nature of developmental phenomena or the interrelationships between genes and other elements of the developmental system. In this regard, recall Monod's final admonition, in the passage already cited, that 'the epigenetic building of a structure is not a *creation*; it is a *revelation*' – in particular, a revelation of what is genomically preformed. I will urge the opposite view in this and subsequent chapters; hence, *creative development*.[7] My account of creative development, to be offered in Chapter 5, represents a substantial modification of all three theses comprising the modern consensus on development. Let us now address these theses in turn.

GENETIC INFORMATIONISM

The tendency toward vitalism has been vigorously challenged by biologists for more than a century (longer, in the case of Aristotle). Again, Mayr is a good exemplar. He characterises himself as an epistemological and methodological emergentist (as I characterise myself in Chapter 1): 'in a structured system, new properties emerge at higher levels of integration which could not have been predicted from a knowledge of the lower-level components . . . Analysis should be continued downward only to the lowest level at which this approach yields relevant new information and insights'. However, Mayr also contends that this emergentist position requires a further concession, one that I am unprepared to grant. This concession is that 'it is the genetic program which controls the development and activities of the organic integrons that emerge at each successively higher level of integration' (Mayr 1997: 19, 20).[8]

In fact, writes Mayr, 'the genetic program is the underlying factor of everything organisms do. It plays a decisive role in laying down the structure of an organism, its development, its functions, and its activities'. For Mayr, then, the genetic programme – 'the information coded in an organism's DNA' – generates the emergent organism (Mayr 1997: 123, 307). Thus does Mayr's account of ontogenesis combine the three theses of genetic informationism, genetic animism, and genetic primacy already identified as comprising the modern consensus. Setting aside until Chapter 4 the question of the primacy of the genes, we must ask: What is genetic information? and What – and where – is the genetic programme?

The usual account of why the notion of specifically genetic information is useful in explaining development is well summarised by Oyama:

> The discovery of DNA and its confirmation of a gene theory that had long been in search of its material agent offered an enormously attractive apparent solution to the puzzle of the origin and perpetuation of living form. A material object housed in every organism, the gene seemed to bridge the gap between inert matter and design; in fact, genetic *information*, by virtue of the meanings of

in-formation as 'shaping' and 'animating', promised to supply just the cognitive and causal functions needed to make a heap of chemicals into a being. (Oyama 1985: 12)

Does genetic information in fact supply these functions? It does – but only at tremendous expense. It is difficult to resist the impression that genetic informationism bleeds into genetic animism, inasmuch as G.C. Williams famously urged that 'a gene is not a DNA molecule; it is the transcribable information coded by the molecule ... the gene is a packet of information, not an object' (Williams 1992: 11). This immaterial information, coded in the medium of DNA, specifies the adult structure of an organism. Similarly, John Maynard Smith describes developmental biology as 'the study of how information in the genome is translated into adult structure', and he underscores that it is 'hard to see where else the information' to build adult structure 'is coming from' (Maynard Smith 2000a: 177, 186).[9] So two core aspects of this perspective are that development is best construed as differential gene activation – that is, as differential triggering of genetic information – and that specifically genetic information undergirds what is for Maynard Smith a strong and irreducibly important distinction between nature and nurture – though nurture is required to trigger nature, nature is primary and necessarily so, given that nature is inherited and nurture is not (Maynard Smith 2000a: 189). In other words, information is properly ascribed only to genes, and so exclusively to genetic causes in ontogeny. What, then, is information?

So many senses of the term 'information' abound in genetic contexts that it is plausible to suggest that 'information' is an instance of what the linguist Uwe Poerksen (1995) calls a *plastic word*. A plastic word is an amorphous but scientifically saturated word stretched so far beyond its appropriate field of applicability that it loses coherent meaning.[10] In case that is an uncharitable interpretation of genetic information, we must ask what end is met by invoking the term 'genetic information' and also whether the term has any explanatory content in contemporary biology.

A number of commentators, including Maynard Smith (2000a, 2000b), Sarkar (1996a, 1996b), and Sterelny and Griffiths (1999), have grappled with this issue. Abstracting from their analyses, we see that two coherent accounts of information are possible: *causal accounts*, stemming from cybernetics and mathematical information theory, and *intentional accounts*, stemming in part from the philosophy of mind. However, although both of these accounts render information-talk in biology comprehensible, neither account permits the further claim that only genes are informational Griffiths and Gray (1994).

According to a causal account of information, information is passed along a channel connecting a sender to a receiver. The receiver contains the information, and the information is about the sender. We know that a channel between two systems exists when we can reliably characterise the state of the sender on the basis of our knowledge of the state of the receiver; that is, when 'the state of one is systematically causally related to the state of the other' (Sterelny and Griffiths 1999: 101).

Channel conditions are the factors that connect two systems at either end of a channel. As Sterelny and Griffiths observe, 'there is a channel between the television studio and the television screen whose channel conditions include the machinery at the studio, the relay stations, the atmospheric conditions, the antennae, and your TV set. So what you see on the read-out device of an instrument causally depends on the state of the source and the states of the channel conditions' (Sterelny and Griffiths 1999: 102). Fundamental to this notion of information is that 'the role of signal source and channel condition can be reversed'; consequently, information is covariation and no more (p. 102; also see Griffiths and Gray 1994).

As a result, sources of developmental information are any and all factors with which development covaries. Again, Sterelny and Griffiths: 'if we hold the other developmental factors constant, genes covary with, and hence carry information about, the phenotype. But if we hold all developmental factors other than (say) nutrient quantity constant, the amount of nutrition available to the organism will also covary with, and hence also carry information about, its phenotype' (1999: 102; also see Sterelny 2000a: 196). Thus developmental information may be found in a wide range of sources, depending on where one chooses to look; genes are not informationally privileged in any non-pragmatic sense on the causal account of information.

Therefore, those who seek in genes a *special* source of information must have another account of information in mind. Indeed, they do. The intentional account of information, as I already noted, stems largely from the philosophy of mind. That an idea has intentional content suggests that the idea is *about* something; the *aboutness* of an idea is its intention (and, possibly, its intension). In the context of genetics and development, the suggestion is that only genes, and not other developmental resources, are *about* the phenotype; the other developmental resources, including cellular conditions and exogenous environmental conditions, help to establish the phenotype but in a completely non-specific way. Sterelny and Griffiths put it this way: 'if genes have intentional content, then they mean the same thing no matter what the state of the rest of the developmental matrix' (1999: 104). Should the matrix change, then the new meaning of the genes amounts to no more than misinterpretation of the original intentional content.

The ideas of misinterpretation and misrepresentation are crucial to the intentional account of information: I misrepresent the colour red if my idea of 'red' is in fact the idea of 'box'; similarly, a map misrepresents the Halifax harbour if the map represents the harbour as on the northern coast of Nova Scotia, and a student essay misrepresents the adventures of Sir Francis Drake if the essayist represents Drake as having circumcised the world with a 100-foot clipper.

So the notion of genetic information having intentional content stipulates that genes are properly about phenotypes; that it makes sense to say that the meaning of genes is relatively independent of other conditions; and that this meaning may be misrepresented in particular environments. Notwithstanding difficulties in the philosophy of mind with the idea of intentional information, it is worth pondering the *source* of the putative aboutness of genes. From whence does meaningful, though misinterpretable, genetic information arise?

Maynard Smith (2000a) holds that the intentionality of genetic information derives from natural selection on the genome in order to generate an adapted organism. This is what is often called a *teleosemantic* theory of informational content, whereby the intended meaning of the content reduces to what it has been selected to be about. Setting aside the outstanding problems of the units of selection and the scope of selection in explaining evolution (which Maynard Smith resolves by appeal to genic selectionism and adaptationism), we see that the core of his position is that genes are uniquely informational because genes uniquely contain intentional information about adult forms, and the intentionality derives from evolutionary history. Maynard Smith's restatement of the core theses of the modern consensus amounts to the following claims, which he takes to be not in dispute:

> that DNA contains information that has been programmed by natural selection; that this information codes for the amino acid sequence of proteins; that, in a much less well understood sense, the DNA and proteins carry instructions, or a program, for the development of the organism; that natural selection of organisms alters the information in the genome; and finally, that genomic information is 'meaningful' in that it generates an organism able to survive in the environment in which selection has acted. (Maynard Smith 2000a: 190)

Whereas genes are thus informational on this view, environments are not; this is because genes are intentional, and environments vary between being supportive and noisy. For Maynard Smith, then, all developmental specificity resides in the informational genome on account of selection for the developmental effects of genes.

This idea of genetic information abounds in experimental contexts. Consider gene-knockout experiments. It certainly makes sense to suggest that the misexpression of the *Otx-2* gene in mice leads to the absence of virtually all structures in the rostral head. The environment in which neither copy of *Otx-2* is expressed differs drastically from that in which one copy is expressed (resulting in a variety of defects – or errors – ranging from minor cranial and neural difficulties to complete acephaly), and this second environment differs drastically, though less so, from that in which the gene is properly expressed. In the latter environment, the rostral brain develops as it should – as it is *meant* to develop.

Of course, sometimes organisms develop as they do – or at least toward a mature form which we could hold to be the form toward which they are meant to develop – sometimes in spite of genes. Knockout experiments are not always as clear cut as in the case of *Otx-2*, and organisms may manage to achieve a typical functional state even in the absence of genes whose putatively intentional effects are well known. The *MyoD* gene in vertebrates plays a key role in determining whether a cell becomes a muscle cell; it is a muscle determination factor. But whether *MyoD* has that effect – and so that intentional content – depends on the prior history of the particular cell and its previous interactions within developmental networks; and even when that gene is misexpressed (as in a knockout experiment conducted by Chen et al. 1992, on its homologue, *hlh-1*, in the nematode worm *Caenorhabditis elegans*), protein synthesis occurs and muscles form as normally expected (for discussion, see Strohman 1993 and Robert 2001a).

However, isn't this just to say that genes are meaningful in meaningful environments, and also that the putative intentionality of organismal development is a matter not just of genes but of genes-in-organismal-context? It is gratuitous to suppose otherwise.

Genes do not evolve independently of the organisms of which they are part, just as environments do not evolve independently of the co-adapted gene–environment complexes inhabiting and altering them. Weiss and Fullerton suggest that we consider 'that it is *not* the genome that is especially conserved by evolution. Suppose the ephemeral phenotype really *is* what we need to understand and what persists over time. Genes would then be "only" the meandering spoor left by the process of evolution by phenotype' (Weiss and Fullerton 2000: 192; see also Newman and Müller 2000 and Robert 2001a). The implication of this view is that we must explore in detail the 'dual evolution of phenotype and genotype' (Weiss and Fullerton 2000: 192) and not presume that what we see in the world around us is strictly the product of genes selected for their putatively intentional informational content. For if genes are in fact so selected, so too are those other aspects of the developmental manifold; and aspects of the whole developmental manifold interact in the process of generating an adapted organism.

Thus, on neither the causal nor the intentional account of information are we entitled to conclude that only genes are informational. Either there is no extant account of uniquely genetic information which is biologically plausible, or there is no such beast as uniquely genetic information (Griffiths & Gray 1994; Sterelny & Griffiths 1999).

And, yet, information-talk persists virtually unabated. Maynard Smith claims that his account is a 'natural history' of this kind of language in biology, not a philosophical analysis (2000a: 192). When a philosophical analysis is undertaken, we see that the goal of elucidating a special informational role for genes cannot be attained. So 'genetic information' is a metaphor – a historically productive one, to be sure, but a metaphor nonetheless. The metaphor masks two implicit claims: first, that even if the metaphor were to remain largely unanalysed, it would continue to do important work; and second, that no specific definition of 'genetic information' is required for 'everyone knows' what is meant by the term – it is used as shorthand for whatever special developmental and evolutionary powers are widely believed to accrue to DNA. Both of these implicit claims may be problematised, though: the metaphor may well have outlived its usefulness in shifting from convenient shorthand to universal explanatory entity, and yet genes do not have the mystical powers attributed to them in the absence of all else that is transmitted between generations and that makes DNA meaningful in specific ontogenetic contexts.

GENETIC ANIMISM

In natural history terms, genetic information has functioned as a surrogate for the assumption that causal control resides in DNA. In a hybrid causal–intentional account of information, François Jacob characterises information as containing 'the power to direct what is done' (1973: 251), even while, in order to function, genes-as-messages require a functioning cell in order to function as guiding causes in development. His argument is that

> outside the cell, without the means to carry out the plans, without the apparatus necessary for copying or translating, it [DNA] remains inert, like a tape outside

the tape recorder. No more than the memory of a computer can the memory of heredity act in isolation. Able to function only within the cell, the genetic message can do nothing by itself. It can only guide what is being done. (Jacob 1973: 278)

Not surprisingly, this idea of genetic information as specifying a programme containing the plans for building an organism fits well within the theses of the modern consensus: genes encode preformed information, which is causally efficacious in controlling the development of the organism, itself nothing but the unfolding of what is genetically (informationally) preordained. In this way, genetic informationism bleeds into the second thesis of the modern consensus, genetic animism.

Recall Mayr's twin claims that a genetic programme is the information encoded in an organism's DNA, and that the genetic programme controls the development and activities of the whole organism. Drawing an analogy with computer programmes is unhelpful in specifying the nature of putative genetic programmes, for a computer programme is such only on account of its relation to the intentional programmer. Oyama similarly worries about the teleological perspective driving the invocation of genetic programmes, the idea that a genetic programme encodes information to direct ontogenesis toward a particular, species-specific goal:

As we contemplate the nature in and around us, the argument from design is ever present. When we remember that our cognitive metaphors are motivated by it, as is the case when we say that an embryo develops as though it had a full set of instructions, all is well. When we forget, we entrap ourselves in the worst kind of pseudoexplanation. What is 'worst' about such explanation is not that it explains nothing, but that it seems to explain everything. (Oyama 2000b: 73)

Nonetheless, as with genetic information, it is worth asking what the concept of a genetic programme is supposed to explain. I am not convinced that it is in any way explanatorily helpful, nor am I convinced that it is philosophically well motivated – despite the fact that the genetic programme trope, like the genetic information trope, has played an important role in explorations of development to date (Keller 2002).

Of particular critical moment is the notion that information *qua* onto-genetic-controller-of-developmental-instruction is supposed to be present *in the genes*. However, as Gray (1992: 177) has argued, in a statement meant to counteract the latent preformationism of the modern consensus, 'develop-mental information is not *in* the genes, nor is it *in* the environment, but rather it develops in the fluid, contingent *relation* between the two'. That genes and environments interact in the production of organisms is not in question; but exactly what such interaction entails is a matter of dispute. My thesis is that proponents of the modern consensus pay lip service to interactionism and then proceed as if genes were the primary generating and determining factor. The task of justifying this assessment will preoccupy us throughout the rest of this chapter and the next.

Some uncontroversial facts about DNA point up the dubiousness of the genetic programme trope.[11] DNA is a relatively inert molecule, requiring activation from without. Further, in eukaryotic cells (cells with nuclei), DNA is covered in histone proteins and therefore is not immediately accessible without cellular triage. The cellular environment which exploits the DNA is complex: even the simplest eukaryotic cells have a ribosomal 'machinery' comprising 'a giant assemblage of sub-units together containing more than 80 different proteins, and RNA sequences containing more than 6,700 nucleotide bases. Without it, without the complex biochemical environment the cell provides, "genes"... simply can't function' (Rose 1997: 127–128). Moreover, gene 'activation' is irreducibly spatiotemporal, depending on the developmental history of the particular cell in which it is located – particularly, the cell's location in the developing embryo and the number of times the cell line that leads to it has divided. Thus, it is evident that genes are not passive providers of encoded instructions that retain their structure across generations; they are 'reactive complexes that are in constant and dynamic interaction with their carriers' (Plotkin 1994: 39; also see Nijhout 1990 and Wolf 1995). In short, in the production of an organism, segments of DNA interact with proteins, metabolites, nutrients, and other segments of DNA according to a specifically structured (though flexible) schedule within a specifically structured (though not invariant) environment which enables such interactions and which is necessary for their occurrence.

Despite the fact that certain of these complex processes can be made to appear to function in a programmatic way, that is not evidence of a genetic programme. At most, as Wolf aptly notes, 'program in this context is an a posteriori description of a structure, and not an a priori instruction for generating a structure' (Wolf 1995: 143; also see Oyama 1985: 54). The ostensibly preformed informational 'instructions' are not 'just there' to begin with but rather emerge progressively during ontogenesis. Thus, Eva Neumann-Held argues that

> independently of context and system, the DNA has neither structure, nor function, nor program, nor information. Rather, the constancy of the transcriptional processes [e.g.] has to be attributed to the constancy of the patterns of interaction

of the participating components . . . The basic DNA sequence and the developmental context determine in reciprocal contingency the structure (and function) of all regulatory sequences of transcription or translation; they *co-define* and *co-construct*. (Neumann-Held 1999: 119)[12]

Thus there is no underlying genetic programme, but at most the illusion of such a programme in the ontogenesis of an organism. (Also see Nijhout 1990: 443.)[13]

Lenny Moss (1992) has subjected Mayr's particular claims about genetic programmes to critical scrutiny. Moss attempts, in vain as it turns out, to localise the genetic programme, reified or treated as a substantial entity by those committed to genetic animism. According to Moss, part of the reason that genetic animists adopt the terminology of genetic programmes is that they draw unjustified inferences from their primary investigative tool: viruses as model organisms. The attribution of agency to genes is facilitated by the dramatic evidence of the formidable effects that the introduction of a virus can have on an organism. The penetration of viral DNA (or RNA) can have a drastic impact on the behaviour of an infected cell, and the observation of this impact may have led investigators to overestimate the agentic role of DNA in ontogenesis. However, as Moss points out,

what becomes easy to overlook in the midst of such apparent power and efficacy is that viruses are molecular parasites whose ability to act entirely presupposes a living system, in relation to which the virus is a kind of trigger or perturbant. Shooting DNA constructs into a cell and shouting 'Now dance!' does not constitute an explanation of the mechanisms by which 'the genetic program informs and instructs ontogeny' or 'supervise[s] its own precise replication and that of other living systems such as organelles, cells, and whole organisms' (even if the cell dances). (Moss 1992: 340–341)[14]

Despite historical and contemporary overstatement of the agency of the genes, Moss is open to the possibility that the notion of a genetic programme may perform actual explanatory work in cell and molecular biology – yet he finds no evidence to this effect. With regard to transcription, for instance, he concludes that the harder one looks for the proximal cause of the transcriptional activation of a particular gene, the quicker one is thrust into a complex array of antecedent conditions – a point made as well by Nijhout:

When we trace the causal pathway of a developmental event, we may often (but not necessarily always) encounter a gene whose product is required for that event, and without which that event would not take place. But the causal pathway does not end there. The expression of the gene or the activity of its product must itself be controlled by a specific stimulus, perhaps an ionic or organic inducing molecule, or through the product of a regulatory gene. Regulatory genes, in turn, owe their timely activity to stimuli external to themselves, and so forth. The causal pathway is endless and involves not only genetic, but manifold structural, chemical, and physiochemical events, a defect in any of which can derail the normal process. (Nijhout 1990: 442)

In the case of cellular self-replication, the three-dimensional structure of a cell is required for DNA to be able to function as a template for the amino acid sequence of proteins; meanwhile, it is not DNA but rather the organelles, variously found in the plasma of a cell, that serve as their own template for replication. As I have already emphasised, without the highly structured cellular environment, which is itself not constructed by the DNA, DNA is inert, relatively unstructured, non-functional, and so not ontogenetically meaningful. Any quest for causal origins (or ontological or ontogenetic primacy) culminates in the intricacies of the cell-organism *as a whole* as the causal basis of 'gene action'. Accordingly, Moss concludes that the 'genetic program upon which Mayr relies so deeply is in fact nowhere to be found' (Moss 1992: 344, 335). Genetic animism is snookered.

Mayr's version of the modern consensus stumbles over the organism-as-a-whole – and its ontological, epistemological, and methodological sequelae. Mayr attempts to argue that the organism is both cause and effect of itself, noting that 'development, behavior, and all other activities of living organisms are in part controlled by genetic (and somatic) programs that are the result of the genetic information accumulated throughout the history of life'; yet he urges that 'it is the genetic program which controls the development and activities' of the organism. He insists that biologists have finally understood that 'the behavior of developing cells [is] attributable not just to genes but also to the cellular environment in which these cells found themselves at different stages in development'; yet he claims that 'the genetic program is the underlying factor of everything organisms do. It plays a decisive role in laying down the structure of an organism, its development, its functions, and its activities' (Mayr 1997: 21, 20, 152, 123).

Well, which is it? Is ontogenesis 'partly controlled' or 'wholly controlled' by a genetic programme? Is the genetic programme 'decisive', or is 'decision making' rather a function of the cellular environment? Note that Mayr invokes genetic programmes to serve two purposes: the first is to explain development, whereas the second is to establish the autonomy of biological science. In the latter instance, he sees two ways to establish his claim that biology is in fact an autonomous science: by distinguishing between life and non-life on the basis of genetic programmes, and by underscoring the unique role of emergence in biology. These two ideas are conjoined in Mayr's pseudo-antireductionistic account of organicism, but, as I shall suggest, the position is inconsistent.

I refer to pseudo-antireductionism for several reasons: first, Mayr is not an antireductionist in the antianalytical sense sometimes implied, for he encourages reduction where appropriate; second, Mayr's position on reductionism is somewhat ambiguous. With regard to methodology, he advises against greedy (methodological) reductionism, insisting that 'analysis should be continued downward only to the lowest level at which this approach yields relevant new information and insights' (Mayr 1997: 20). On the other hand, Mayr falls prey to a certain genetic determinism in his endorsement of an ontogenetically primary and decisive genetic programme. Mayr contends that the recognition of emergence tempers the apparent reductionism and determinism of a focus on genetic programmes; I harbour doubts.

The basic features of his distinction between animate and inanimate matter are the genetic programme (*qua* depository of developmental information and director of ontogenesis), and the presence in animate matter of properties emerging from the particular organisation (structure) of animate matter. Only living beings contain and are generated by genetic programmes – hence the (ontological) autonomy of the biological subject matter. Given the emergent properties of living beings, which imply the inexplicability of (some) higher-level properties from knowledge of lower-level properties, biology must be an (epistemologically) autonomous science. That, in a nutshell, is Mayr's organicist–structuralist position on the autonomy of biology.

As noted in Chapter 1, as a systemist I contend that an organism is an emergent outcome of its composition, environment, and structure. In contrast, Mayr ignores environment altogether and denies the importance of composition: 'the unique characteristics of living organisms are not due to their composition but rather to their organization'. Yet as is plainly evident, Mayr also holds that the key compositional feature of an organism is its informational genetic programme, which is responsible for the generation of 'integrons' at all higher levels of organisation throughout development (Mayr 1997: 16, 19–20). How can he reconcile these views?

In brief, he can't. Moreover, Mayr neglects to clarify the ontological status of what he argues is the basic compositional element of organisms, that is, their genetic programme. A genetic programme is, apparently, not immaterial, for Mayr claims that vitalism is superfluous; but whether it is strictly physical is also unclear, for Mayr rejects physicalism as well (Mayr 1997: 8). He does not clarify which particular meaning of 'information' he endorses in claiming that a genetic programme just is the information encoded in an organism's DNA. Nor does Mayr adequately justify the epistemological component of his account of emergence: if a genetic programme encodes and directs the organisation of the organism, and the higher-level properties are 'emergent' from the genetically preprogrammed organisation of lower-level parts, then it is at least not obvious why Mayr should contend that higher-level properties should be either inexplicable or unpredictable from lower-level organisational properties.

I submit that the way Mayr or someone attracted to his position might avoid all of these problems is to insist on genuine (ontological) emergence; to exchange structural organicism for systemism; to drop the very notion of a genetic programme; and to insist that whatever seeming inexplicability or unpredictability exists is an artifact of too quick and exclusive a focus on genetic information as generating the explanation or prediction in the first place.

NAVIGATING THE STRAIT

Melding preformation and epigenesis, Scylla and Charybdis, is the wrong project. The chimaeric offspring is even more beastly than its monstrous forebears. In various versions of the modern consensus, differential states of activity are attributed to both preformed and epigenetic elements. For the Medawars, for instance, genes encode developmental instructions that are triggered from outside the cell, suggesting relatively passive (preformed) genes in a relatively active (epigenetic) environment; for Mayr, though, encoded instructions in genes simply *are* in themselves a genetic programme for development (that is, in fact, his definition of a genetic programme) – suggesting a rather active role for the preformed component.

I have already suggested (and will continue to do so) both that genes are much less active than sometimes presumed and also that genes, *qua* passive suppliers of (some of) the materials of ontogeny, 'propose' by no means all but rather only some of those materials to be 'disposed of' epigenetically. Having worried the first two theses of the modern consensus herein, I turn in Chapter 4 to the third thesis: genetic primacy. In Chapter 5, I endorse an account of development distinct from and not reducible to the differential expression of (preformed) genes.

Admittedly, I have offered here some harsh criticisms of a number of great biologists. Some may object that I have been too severe, or have constructed a straw version of sophisticated accounts of the role of genes in explaining development: belittling great biologists is philosophically unconstructive, whereas toppling a straw consensus is philosophically unimpressive. In response, I emphasise that the modern consensus is indeed widely believed, though in various incarnations, and that my criticisms of these views are not ends in themselves. They are, rather, part of the larger project of genuinely understanding and explaining development as such – and so navigating safely between Scylla and Charybdis.

4

Constitutive Epigenetics

Developmental biologists and geneticists usually focus on different aspects of genes (translation versus transmission). The geneticist uses a particular view of genes as units of heredity (i.e. transmission to the next generation) and may neglect the role of genes in development. Consequently, the developmental biologist may ask whether the distinction between genotype and phenotype advances genetics by leaving out development. Does evolutionary genetics provide a sufficient theory of morphological evolution? The mapping function from genotype to phenotype is not one-to-one. A gene may affect multiple structures (pleiotropy) and traits are often affected by many genes (polygeny). Furthermore, the mapping of gene effects on phenotype may be nonlinear. Because gene action during development is a cyclic series of gene-cell interactions, genes are just one element in the developmental process. Thus the nature of interactions is the primary issue in development.

– S.J. Arnold et al. (1989)

Let us briefly take stock. In Chapter 1, I suggested that the problem of development should not be reduced to the problem of gene action and activation. In Chapter 2, I provided three examples to orient the discussion throughout the rest of the book, the first two of which have been implicated in the argument so far. In Chapter 3, I introduced and explored the nature of the modern consensus on development, according to which development is construed as a matter of the epigenetic activation of preformed genetic information. I then investigated this putative reconciliation of epigenesis and preformation, and particularly the twin theses of genetic informationism and genetic animism (which will again concern us in this chapter). I contended, following numerous others, that no coherent account of biological information has yet emerged that would justify the usual position that genes are uniquely informational.

Moreover, I suggested that the metaphorical use of 'genetic information' as programming development instantiates a curious resurgence of vitalism. Note the historical peculiarity of this last claim: most preformationists up through the nineteenth century were materialists focusing on the growth of preformed matter; most epigenecists had immaterialist tendencies toward vital forces guiding material development. Within the modern consensus, we witness a reversal: the immaterial genetic programme is preformed, whereas epigenesis reflects the material emergence of organic complexity.

My argument is that we should not attempt to meld together these two unsatisfactory ideas, preformation and epigenesis, in the ways explored in Chapter 3. Although they each may contain a kernel of truth, each also carries a lot of historical and ideological baggage. We should, instead, find the golden mean between them, not by bringing them closer together but rather by clearing enough conceptual room to avoid both of them.

Throughout the discussion thus far, I have implicitly worried aspects of the third thesis of the modern consensus: genetic primacy. In this chapter, I address it directly. I should note at the outset that the thesis of genetic primacy has an odd relationship with the interactionist consensus. It is clear that genes are not self-governing, for 'their expression – whether they are active or inactive – is determined by influences from other levels of the system' (Gottlieb 1995: 138). It is clear, as well, that 'the reductionistic-analytical approach has had great success in detecting and understanding genes at the molecular level but is far from adequate for understanding how a gene functions in the larger context, where there is a web of feedback relations, nonlinear interactions, and multifactorial contingencies' (Wahlsten and Gottlieb 1997: 179). Both of these truisms have been incorporated into the interactionist consensus, according to which both genes and environments (on many scales) are required for the production of phenotypes. But, I shall argue, the significance of these truisms has yet to be fully recognised, which helps to explain why those who hold to genetic primacy also often subscribe to the interactionist consensus. My claim is that this is because we too easily fall into the trap of interpreting development in additive terms – what I call a genes-*plus* interpretation of development: a standard (or, for that matter, non-standard) environment triggers or activates a particular ontogenetic response from the genes, leading to a particular phenotypic outcome. Though we now recognise that these 'triggers' exist at a variety of levels, and we are beginning to explore how gene function is regulated, we have yet to digest the fact that gene structure is not immutable but rather that genes are assembled on the spot in response to the needs of the organism-in-its-surround and also that 'triggering' is therefore not an adequate explanation of developmental interactionism. It is the burden of this chapter to elaborate and justify these claims.

The primacy of the gene has been with us for a century (Keller 2000). The history of the conceptual and methodological split between genetics and embryology at the beginning of the twentieth century has been told often, and much better than I could tell it. Nevertheless, it is worthwhile to take a brief detour through some of this history in order to appropriately grapple with the primacy of the gene.

EMBRYOLOGY *AND* GENETICS; EMBRYOLOGY *AS* GENETICS

One significant foe vanquished in the early days of 'the rise of genetics' (Morgan 1932a, 1932b) is the discipline of embryology. By the 1930s, only embryologists took embryology seriously, while most others were attuned to the elegance, austerity, and simple beauty of genetic models which reduced ontogeny to gene activity. Thanks to its tremendously productive models, genetics established a monopoly position in both evolutionary biology and embryology in the first half of the twentieth century: embryology was redefined as the study of changes in gene expression, whereas the task of evolutionary biologists was recast as the study of changes in gene frequencies in a population (Gilbert et al. 1996: 360; Morgan 1934). Embryologists were not corporately impressed with the theories of the geneticists, though, especially in the developmental realm.

Jan Sapp notes that 'American geneticists repeatedly stated that one day they would be able to account for development in terms of the governmental control of chromosomal genes' (1991: 237). However, as N.J. Berrill cheekily remarked (Gilbert et al. 1996: 361, citing Berrill 1941), genes were no more than 'statistically significant little devils collectively equivalent to one entelechy'! Nevertheless, embryologists feared the takeover of their discipline by 'marauding intruders', as Berrill referred to the geneticists of the 1930s.[1] As early as 1924, Hans Spemann remarked that the geneticists' 'previous progress has been amazing, and it is not from a feeling of futile labours but rather from being aware of their paramount powers of appropriation that geneticists now are on the look-out for new connexions. They have cast their eye on us, on *Entwicklungsmechanik* [developmental mechanics]'.[2] Ross Harrison, among others, warned against the geneticists' *Wanderlust*, their unwelcome intrusion into the developmental realm; his concern was that the successes of genetics in the domain of transmission would lead to an overemphasis on the genes in the domain of development (Harrison 1937). Harrison's counsel was, as we shall see, prophetic, and therefore unfortunately impotent.

Some geneticists had mixed feelings about embryology. Thomas Hunt Morgan was himself an embryologist, and one not easily won over to the geneticists' side. Between 1900, the date of the rediscovery of Mendel's work, and about 1910, Morgan was critical of Mendelian theories of heredity on the basis of their alleged adherence to August Weismann's version of preformationism. In his 1892 book *Das Keimplasm* (*The Germ Plasm*), Weismann explained that he had earlier rejected the preformationist notion of the uncoiling of a tiny preformed organism, but now he had found that it was impossible to accept anything other than *some* kind of preformationism. He was therefore 'forced into accepting' the position he once resisted (Maienschein 1986: 75–76), a position he shared with Wilhelm Roux.

For Weismann, what was preformed was neither the organism nor 'the organs in miniature, but organic particles corresponding to *and determining the growth of* the organs' (Russell 1930: 31). Weismann identified these organic particles as the 'idioplasm', by which he meant the chromatin granules in the nucleus. The idioplasm 'exercises a direct formative influence upon the cell containing it, determining what sort of cell it will become ... The idioplasm of the germ-cell – the hereditary or germinal substance proper [for Weismann] – is conceived to be of a complex and orderly architecture, built up of self-propagating units or determinants, each of which is destined to be the formative agent of some particular part of the organism or of some particular group of cells' (Russell 1930: 43). For Weismann, then, factors external to the nucleus were specifically irrelevant – they were simply the normal conditions for development – whereas internal factors alone determined ontogenesis. Weismann summarised his position as follows: 'a certain cell in a subsequent embryonic stage does not give rise to a nerve-, and muscle-, or an epithelial-cell because it happens to be so situated as to be influenced by certain other cells in one way or another, but because it contains [in its nucleus] special determinants for nerve-, muscle-, or epithelial-cells' (Weismann 1893: 134).

Morgan, at first, found these 'special determinants' to be troublesome. For if we identify hereditary factors with characters, then we ignore every ontogenetic process between gene and phene: that is, exactly those phenomena that have long since fascinated and baffled embryologists. In a passage reminiscent of both Bonnet and Weismann, the geneticist L.C. Dunn attempted to clarify this notion of genetic (pre)determination by claiming that the use of the word *determined* 'does not mean that the character itself is present in the germ in any form, but rather that it is represented by substances or forces which not only *stand for* the character but in some way bring about its expression' (Dunn 1917: 286). The idea is that there is some direct connection, some linear path, between the factors and the characters. However, without any specification of the details of the ontogenesis of the character from the factor, the theory was thought to be incomplete.

Morgan noted in 1926 that there was a gap needing to be somehow filled: 'between the characters, that furnish the data for the theory, and the postulated genes, to which the characters are referred, lies the whole field of embryonic development' (Morgan 1926: 26). By this time, Morgan-the-geneticist no longer found this lacuna to be problematic. However, Morgan-the-embryologist, at least before 1910, could not ignore this ontogenetic gap, and therefore he could accept neither Mendelism nor any chromosomal theory of heredity.

The reason is that the embryologist Morgan would not, could not, separate heredity from development: 'learning about transmission of information between parents and offspring was of no value without also learning about the development of the trait into its ultimate adult form' (Allen 1986: 120). The geneticist Morgan redefined heredity as transmission and no longer also as ontogeny. That left for embryology the study of development, and for genetics the study of heredity (or transmission). Thus, for the Morgan school of transmission genetics, heredity did not refer to the development and reproduction of individual organisms but only to the sexual transmission of their genes. That the gene theory said little about ontogeny was no longer a problem for Morgan thanks to the new, restricted (and remarkably productive) understanding of heredity.

Garland Allen has plausibly argued that Morgan bracketed ontogeny for pragmatic reasons. Embryonic development was knotty and messy; transmission was clear and straightforward, and experimental success with *Drosophila* in showing any number of genetic alterations was practically guaranteed. The distinction between genotype and phenotype, introduced in 1909 by Johannsen, facilitated this disciplinary and experimental wedge between genetics and embryology, and though Morgan remained interested in embryology, he felt a pressing need to push it aside in favour of transmission genetics (Allen 1986: 126–127, 138–139). Nevertheless, Morgan himself sometimes slipped, urging for instance in 1919 that 'one could account for "the organism as a whole" in terms of "the collective interaction of genes" ' (Morgan 1919: 241). Therefore, Morgan did have a theory of development: gene expression. It just wasn't a very good one.

In sum, though Morgan kept the disciplines of genetics and embryology officially separate, securing research funding for the former and helping to establish its central position in American biology, he nevertheless occasionally insisted that development could, and should, be explained genetically. Morgan's 1926 sense that 'the application of genetics was a most promising method of attack on the problems of development' was widely held (Sapp 1987: 50, referring to Morgan 1926: 491–496). Although he was at times 'particularly anxious' to dismiss claims about the importance of extra-nucleic factors (cited by Allen 1986: 131, from a letter from Morgan to Jacques Loeb in 1931), Morgan nonetheless proposed in 1934 a more expansive concept of the gene than was then commonplace – one that kept up his early interest in the nongenetic elements of the ontogeny of organisms.

Gilbert notes that on the last page of Morgan's 1934 book *Embryology and Genetics*, Morgan 'suggest[ed] that the nuclear genes may not be the unchangeable entities that geneticists had (and until very recently still have) assumed'. Morgan wrote that it is 'conceivable that the genes also are building up more and more, or are changing in some way, as development proceeds in response to that part of the protoplasm in which they come to lie, and that these changes have a reciprocal influence on the protoplasm' (Gilbert 1988: 315, citing Morgan 1934: 234). In a sense, then, Morgan was more generous in keeping embryology and genetics separate and retaining (at least some of the time) appropriate regard for both disciplines than were those who, unlike Morgan, have since the 1930s attempted to reconcile genetic and embryological research programmes. I have in mind early proponents of such a synthesis, such as Richard Goldschmidt and Ernest Everett Just. (For an extended discussion of their proposals, see Gilbert 1988.)

GENETIC PRIMACY

Though there have been numerous, sometimes high-profile, detractors,[3] research agendas in biology over the past ninety years have tended to converge on a genetic account of ontogeny, one according to which the nucleus holds court over the rest of the cytoplasm; as the Austrian physicist Erwin Schrödinger pithily and presciently remarked in 1944, the genes are well understood as 'law-code and executive power – or, to use another simile, they are architect's plan and builder's craft – in one' (Schrödinger 1944: 23). Almost sixty years have passed since Schrödinger published *What Is Life?* and, ever since, this picture of the gene has become further deeply entrenched in biological research and ideology.

As indicated in Chapter 3, those who believe in the thesis of genetic primacy envision the gene as the unit of heredity, the ontogenetic prime mover, and the primary supplier and organiser of material resources for development. Accordingly, the phenotype is the secondary unfolding of what is largely determined by the genes. Those who insist on genetic primacy in conjunction with the interactionist consensus do not insist that genes are everything: genes must be activated and regulated epigenetically in order to have their developmental effects, but genes are nonetheless primary, for they carry all relevant (specific, intentional) developmental information. Genetic primacy should not be mistaken for genetic determinism, though. Genes alone are not sufficient for phenotypic traits; given the interactionist consensus, genes interact with other (secondary) developmental resources during ontogeny to produce the phenotype. But what is the nature of interaction?[4]

Interaction may occur at both populational (analysis of variance) and individual–developmental (analysis of causes) levels. At the population level, the task is to explain differences in traits in a population; that is, the task is to account for phenotypic variation in terms of environmental variation, genetic variation, or both. From this perspective, interaction may be understood in two ways: additively (genes + environment = phenotype) or non-additively (genes × environment = phenotype). Additivity in this context refers to the aggregation of independent influences – the contribution of the genotype is insensitive to any environmental factor, and the contribution of the environment is not influenced by the genotype (Lewontin 1974; Wahlsten and Gottlieb 1997). There is a longstanding dispute between those who downplay non-additive interaction and emphasise additivity and those who recognise significant non-additive interaction.

In the context of individual development, the task is to explain the source not of differences in traits but rather of the traits themselves. It is thus the effort to understand the causal activities of genes and environments in the ontogenesis of a trait. Here, the situation is equally charged. A significant source of disagreement in this domain involves conceptual slippage from the level of populations to the level of individual organisms. Though there is an obvious difference between understanding statistical variance-in-traits and understanding ontological causes-of-traits, the two have sometimes been confused, and so the causation of traits has been partitioned into genetic and non-genetic components just as differences in traits have been partitioned (Sarkar 1998).

In the developmental case, interactionism sometimes refers to a thesis about genes (primary) and environments (secondary) as relatively independent factors, whereby genes are environmentally activated to produce the phenotype from what is thought to be latent in the genotype. This is, I take it, the standard view. At other times, though, ontogenetic interactionism is seen as more complex, consisting in a broader range of comparably important inherited and non-inherited factors (DNA, cytoplasmic characters, nutrients, and more), characterised by their context sensitivity (e.g., to temperature), developmental history, and spatiotemporal positioning in the cell and in the organism, multiply influencing each other in the constitution of the phenotype (which is not in fact presupposed in the genotype).

On the former account, the one allied with genetic primacy, interaction amounts to ontogenetically specific-information-bearing genes being expressed as a result of (usually non-specific) non-genetic 'triggering'; genes are primary, though requiring activation for ontogenesis to take place, and the phenotype is only a proxy for the foundational genes. This is what I call a genes-*plus* account of interaction: genes (primary) + environmental trigger (secondary) = phenotype. On the latter account, the one I prefer, the developmentally specific information resides not in the genes but rather in the spatiotemporally delimited developing system, which is therefore the ontogenetically primary unit; accordingly, interaction is not limited to gene activation but rather implicates positive and negative feedback loops at a variety of levels within and without the developing system and which contribute to the very constitution of the organism. Note that there is a qualitative difference between these two alternatives: the latter is not a more detailed presentation of what is implicit in the former but is rather a rejection of its basic premise of genetic primacy.

EPIGENETICS

In recent years, there has been a resurgence of interest in the notion of 'epigenetics'. Conrad Hal Waddington first coined the word 'epigenetics' in 1940. Waddington sought to marry the classical notion of epigenesis (that ontogenesis, whatever else it may be, is not merely the growth of preformed miniatures or potential structures) to the discipline of genetics. René Thom notes that 'epigenetics' has managed well as a concept, not least because of those experimenters and theorists who resist the by-now common sense that the development of organisms is somehow coded in the genes. Such critics were 'inclined to emphasise the importance of local morphogenetic factors such as mechanical strains on tissues (following the *Entwicklungsmechanik* of Roux), the contact with nearby tissues, local environmental influences like external gradients, etc.; hence the need for a new word subsuming all these local events' (Thom 1989: 2–3). However, as even a cursory review of the recent literature on epigenetics attests, there is considerable variation in the meaning of the word.[5]

Even Waddington meant different things at different times. Judging from his early work, Waddington held that epigenetics refers to the causal analysis of development (Waddington 1952; Hall 1992a); such an account obviously overlaps with Roux's perspective on *Entwicklungsmechanik* and was captured more recently by Løvtrup as 'the study of the mechanisms responsible for the effectuation of ontogenetic development' (Løvtrup 1988: 189). However, writing in 1975, Waddington offered a more specific definition, whereby epigenetics denotes the 'causal interactions between genes and their products which bring the phenotype into being' (Waddington 1975: 218).

Although this latter definition is more specific, it is also more narrow; the earlier definition does not restrict itself to genes and gene products and so allows for consideration of non-genetic (e.g., cytoplasmic, hormonal, or positional) factors within the purview of epigenetics. The causal analysis of development, I maintain, is not the exclusive domain of developmental genetics, and the category of developmental interactions is not populated exclusively by gene-based interactions. Accordingly, we should prefer Waddington's earlier account of epigenetics.

In much of the recent literature, though, the narrower definition prevails. Consider this passage from Henikoff and Matzke (1997) in their introduction to an issue of *Trends in Genetics* devoted to epigenetics:

> The term 'epigenetics' was introduced by Conrad Waddington to describe changes in gene expression during development. Nowadays, epigenetics in the Waddington sense refers to alterations in gene expression without a change in nucleotide sequence. However, this definition is so broad that an issue in *Trends in Genetics* devoted to epigenetics would read more like a modern biology textbook than a series of critical reviews. A more focused description of epigenetics refers to *modifications in gene expression that are brought about by heritable, but potentially reversible, changes in chromatin structure and/or DNA methylation.* (Henikoff and Matzke 1997: 293)[6]

Although this account of epigenetics is akin to a genes-*plus* account, note four aspects of this now-common depiction: epigenetics refers to the regulation of gene expression; the regulatory mechanisms are inherited; the regulatory mechanisms are relatively independent of the DNA sequence; and the regulatory effects may be modulated or reversed. The basic idea is that genes are not ready *tout court* to be expressed; whatever message they contain must be accessed through the efforts of various heritable, non-genetic (epigenetic) regulatory mechanisms. This suggests that gene activation and regulation are crucial to development but that genes are nonetheless primary in development.

In their book on epigenetic inheritance systems, Jablonka and Lamb (1995) propose the phrase 'the phenotype of the gene' to account for the context-dependency of DNA transcription. Heritable phenotypic features of genes – for example, methylation patterns, chromatin structure, and genetic imprinting – comprise the epigenetic response to the question 'how do disparate cells containing the same complement of DNA, and the same cell at a different spatiotemporal location, express differentially in time and space in the developing organism?' (Wolffe 1998: 1). I will address each of these epigenetic phenomena briefly in turn.

Much controversy has attended to the function of DNA methylation since its discovery in the mid-1970s (Henikoff and Matzke 1997: 294). Methylation involves adding a methyl group to some of the cytosine residues of DNA, thereby forming 5-methylcytosine, which influences the transcription of the DNA: highly methylated DNA is transcriptionally less active than either unmethylated or less methylated DNA (Hall 1999: 118). It is plausible to suggest that methylation helps to determine the segregation of parts of the genome into inactive and active compartments. Appropriate compartmentalisation is absolutely crucial to gene expression, for the DNA sequence itself is ontogenetically relatively uninformative (Wolffe 1998).

Another aspect of the phenotype of the gene is generated by the structural conformation of the chromatin. Chromatin structure has a 'dynamic nature' that may be modified by genetic restructuring during gametogenesis, for instance, or by genomic imprinting:

> Chromatin is a dynamic complex of DNA, RNA, histone, and non-histone proteins embedded within the eukaryotic nucleus and nuclear matrix. The nuclear matrix is thought to provide the spatial arrangement and the structural framework needed for DNA replication, transcription, recombination, and nuclear transport. During mitosis, chromatin and the supporting nuclear matrix are efficiently disassembled, partitioned, and subsequently reassembled into daughter nuclei. (Riggs and Porter 1996: 39)

One leading suggestion is that the structure of the chromatin plays a role in rendering most of the DNA in a cell off-limits to the transcriptional machinery, such that only the requisite segment of DNA is transcribed at a given time and place (Wolffe, 1998).[7] 'Alteration or modification of chromatin-related structural proteins may provide a dominant means of controlling the transcriptional activity of individual genes, domains, and entire chromosomes' (Riggs and Porter 1996: 40).

65

A third epigenetic influence is the imprinting of the genome at the level of single genes (Peterson and Sapienza 1993). 'In genomic imprinting, two copies of a gene (either maternal versus paternal, or one allele) do not function equivalently during development' (Hall 1999: 118). More specifically, in the case of the differential functioning of alleles depending on parental origin, whether a heterozygote for a particular mutation manifests the phenotype in question is partially contingent upon which parent transmitted the mutant allele.

This is the take-home lesson of Jablonka and Lamb's study of epigenetic inheritance systems: 'many evolutionary and developmental phenomena, which appear puzzling or anomalous on the received view that the origin of all variation traces ultimately to changes in nucleotide sequences, can be understood as having an epigenetic, rather than a genetic, basis' (Griesemer 1998: 107, summarising the conclusions of Jaoblonka and Lamb 1995; also see Petronis 2001). On this view, then, epigenetics is primary.

These three sources of epigenetic influence during development do not operate in isolation. Genomic imprinting may affect the structural conformation of the chromatin, and DNA methylation may be involved in genomic imprinting or in gene inactivation (Hall 1999: 118–119). The deep context dependency of gene expression generated by such epigenetic effects prompts Wolffe to suggest a kind of epigenetic systemism: 'for the propagation of a state of gene activity it is necessary to replicate not only the DNA sequence, but also to duplicate the chromosome and to recruit a gene to the appropriate nuclear compartment' (Wolffe 1998: 3). In other words, 'the epigenetic mark on gene expression [results in] the difficulty of recapitulating the correct control of gene expression without the appropriate developmental history or chromosomal context' (Wolffe 1998: 1).

Immediately, then, one recognises an important constraint on the image of development promulgated by adherents to the modern consensus: not just *any* 'supportive environment' will do for the proper regulation of genes. A very particular environment, one laden with details of spatiotemporal developmental context and cellular memory, is prerequired for genes to make phenotypic sense. We should go still further: 'the nature of the phenotype of any organism cannot be mechanistically deduced, even if we possess a complete DNA sequence of its genome' (Schlichting and Pigliucci 1998: 27). For *contra* Monod, development does indeed require *specific information beyond the genome*: epigenetic information. To understand the relationship between genotype and phenotype, we must transcend the dichotomy between them in two ways: we must grasp the phenotype of the gene and we must recognise that the relevant developmental space does not begin nor does it end with the genome-in-context. It begins, instead, with the genetically *co*-defined primary, initially unicellular, organism: the cell, the zygote, the embryo; and it ends with the developed adult organism, which itself continues developing.

ORGANISM-CENTRED BIOLOGY

In Chapter 3, I disputed the notion of a genetic programme, and with it the thesis of genetic animism. Now, in further disputing the thesis of genetic primacy in favour of epigenetic (developmental) primacy, let us return again to the programme metaphor. In her recent work, Evelyn Fox Keller (2000, 2001, 2002) underscores a crucial distinction between genetic programme and developmental programme. Both concepts, borrowed directly from computer science, came into circulation in the 1960s. Keller notes that the idea of a developmental programme was elucidated by Michael Apter in 1966 but then faded into obscurity, overtaken by the idea of a genetic programme (Keller 2001: 303).[8] Whereas Keller urges the importance of the former (the notion of a programme dispersed throughout the zygote), she is highly critical of the latter (the notion of a programme located in the genome). Thus, although she agrees with Lenny Moss that a genetic programme is nowhere to be found, she extends his account (Keller 2001: 310, note 14) by suggesting that a developmental programme is, by contrast, 'everywhere to be found'!

For Keller, the very idea of a genetic programme rests on the conflation of two independent distinctions: between programme and data, and between genetic and epigenetic (Keller 2001: 302). (Epigenetic here should be understood as non-genetic ontogenetic resource or process.) The result is the association of genes with programmatic agency, and the association of everything-else-ontogenetic with relative passivity. The idea that a genetic programme explains development can be understood in part as a product of its time: development begins with the fertilisation of the (inactive) egg by the (active) sperm; the cytoplasm was then thought of as inactive, implying that the active component of development must be the almost wholly nuclear sperm; so the genetic information in the nucleus must contain the programme for the sequential activity of genes in development.

However, there is a difficulty with, for instance, François Jacob's assertion that a genetic programme equates the genetic material with a computer's magnetic tape. The metaphor is quite optional, even gratuitous. As Keller notes, genetic material 'might just as well be thought of as encoding "data" to be processed by a cellular "program". Or by a program residing in the machinery of transcription and translation complexes. Or by extra-nucleic chromatin structures in the nucleus' (Keller 2001: 303). Though it may be historically understandable why some thinkers were tempted to talk this way, attributing agency to genes is ontologically and ontogenetically misguided. Thus, Keller prefers the alternative image of a developmental programme dispersed throughout the cell-organism, according to which the genome is not programmatic but rather provides (some of the) data for the developmental programme.

Embryologists in the early part of the twentieth century emphasised the importance of the organism-as-a-whole. As Jan Sapp (1987: 7) has observed, 'the notion that the whole organism subsisted only by means of reciprocal action of the single elementary parts was for them inadequate to explain the harmonious whole manifested by the organism. The fact that each of the parts of the egg was capable of developing into a complete organism, and yet did not do so when left in its natural position, proved that the developing germ, the embryo, was an integrated unit'. For whole-organism biologists, such as E.S. Russell and those whose works he discusses, this is a fundamental fact of biology. Partly because Russell's writings on development have by and large been ignored by biologists and philosophers of biology,[9] and partly because Russell's views foreshadow those of Keller (2001), I will outline Russell's position primarily as elucidated in his 1930 gem, *The Interpretation of Development and Heredity*.

Russell's credo is that 'the organism develops essentially as a whole, as a unitary individual, persisting in time' (Russell 1930: 6). He identifies the germ-plasm of the gene theorists as a problematic 'material entelechy':

> The germ-plasm is, as it were, a material entelechy. The attempt to find an internal formative mechanism as the cause alike of heredity and development, which is characteristic of nearly all modern theories, results necessarily in [the] separation of agent and material, just as the attempt of the vitalists to reintroduce life into the mechanistic abstraction that stands for organism results in a dualism or opposition between the immaterial agent and the material mechanism which it in some way controls. In either case one arrives at a *Deus in machina*. The nuclear organization, the germ-plasm, or the gene-complex of modern theories, is accordingly invested with semi-magical powers of control. (Russell 1930: 154)[10]

Russell seeks to distinguish himself, both metaphysically and methodologically, from the progenitors of the genetic programme trope (and from certain older vitalists, such as Driesch). With regard to methodology, Russell believes that the (ontological) unity of the organism is 'not decomposable without loss' (Russell 1930: 146):

> To regard any process or structure by itself without relating it to the general activity of the organism is to deal with something which is in large measure abstract and unreal. To re-invest it with some degree of concrete reality it is necessary to re-integrate it into the whole. Its isolation by analysis should be provisional only, and after analysis there should always follow re-integration. We know that the reconstitution of the original unity will be incomplete, but we must make it as complete as possible. (Russell 1930: 147)

Yet, Russell contends, too many biologists fail to recognise the limitations of analysis and fail to follow any sort of reintegrative strategy (Russell 1933: 155).

Eva Neumann-Held, for one, has recently spoken to the persistence of such a failure, distinguishing between the 'differentiative' and 'integrative' aspects of scientific descriptions and explanations (Neumann-Held 1999: 106–107). Differentiation is crucial in order to access biological structures or processes, but analysis is not enough; by itself, it degenerates into mere fragmentation, offering no comprehension of the interactive and interreactive relationships among elements of the system of which they are an integral part. Working out these relationships is the focus of the integrative element of science. Neumann-Held concludes that 'in the description of organisms (more generally: of systems), biology still has to perform the integrative part. So far, biology can describe organisms down to the molecular level of genes. However, the interactions of genes with other, non-genetic components to form an organism is far from being understood' (p. 107), as are interactions at many other hierarchical levels.

In order to address this integrative task, Russell proposes two 'cardinal principles' of biological method: (1) *'The activity of the whole cannot be fully explained in terms of the activities of the parts isolated by analysis, and it can be the less explained the more abstract are the parts distinguished'*; and (2) *'No part of any living unity and no single process of any complex organic activity can be fully understood in isolation from the structure and activities of the organism as a whole'* (Russell 1930: 146–147). These two precepts capture the epistemological and methodological elements of systemism without pitching us into the analytical void of holism. As for his ontological commitments, Russell argues that

> there is a unity of the whole organism – it develops as a whole, and acts as a whole – and this unity is not a secondary or composite thing, but primary and original. To distinguish cells as independent unities, having their own modes

of action independent of the action of the whole, is to regard them abstractly, and to introduce an artificial simplification... The ovum and the embryo are from the very beginning unitary organisms... [T]he unity of the organism is not something which comes to be during the course of development, but is there *ab initio*. (Russell 1930: 234–235)

Thus, as Neumann-Held contends, though methodological integration is (or ought to be) of concern to biologists, ontological integration is, as Russell says, 'not a problem for biology':

> If we reject, as I think we must, any vitalistic interpretation in terms of an entelechy or other organizing agent, we have no alternative but to accept the observed facts of development and make the best of them. It follows that the unity of the organism, which is there at the beginning, must be accepted as fundamental; unity or integration is not a problem for biology, but an axiom, a master-fact to which we must relate all other facts about the organism. (Russell 1933: 155)

That this relation between the unified whole and its parts is not merely methodological or epistemological (having to do with ease of investigation or parsimony of explanation) is evident in Russell's further insistence that 'integrative or "whole" action means that the activities of the parts are subordinated to the activity of the whole' (Russell 1930: 232).

To elucidate this latter point, he invokes the idea of the *autonomy* of the developing organism, 'its relative independence of environment, its self-containedness, its steady persistence towards the goal of the finished form':

> The developing organism acts *as if* it were fulfilling an end or purpose – that of arriving at the typical form and modes of activity of the species; it tends towards this goal in spite of difficulties, and the end is more constant than the way of attaining it. The environment supplies the conditions for development, provides the means, and also acts as a limiting factor, but the developing organism reaches its definitive form as it were in spite of environment, utilizing environment where it can, and seeking other conditions when the environment becomes unfavourable to its development... Alteration of environmental conditions [excepting the absence of essential environmental factors] will not produce an essentially different embryo. (Russell 1930: 6–7)

In this passage, we see Russell's signature emphasis on the organism's remarkable ability to self-regulate (the observation of which sent Driesch beyond regulative development toward his eventual preoccupation with metaphysical vitalism).

For Russell, 'if the conditions do not permit of a straightforward normal development, if for instance the developing organism suffers deformation or loss of parts, it has to a considerable degree the power of so modifying the course of its development as to cope with the unusual situation, replacing, for example, the missing parts' (Russell 1930: 7). In other words, it is characteristic of life, or of an organism (in contrast to a machine), to find some other way to achieve its species-typical form of organisation.

Russell notes additionally that 'this typical form is an amazingly exact replica of the form of its parent or parents', and he calls this 'the fact of *heredity*' itself. As such, 'repetition of type must be regarded as one of the main characteristics of development', leading Russell finally to 'treat of heredity as being primarily a feature of development' – without any need for genetic programmes (Russell 1930: 7, 8).

Two crucial implications of Russell's position relate directly to our concerns. First, though he is by no means a supporter of the (nuclear) gene theory of development, neither does his position support any preformationist cytoplasmic developmental theory. His insistence on the original unity of the zygote (the fertilised egg comprising both nucleus and cytoplasm) leads him to recognise 'obviously complex, intimate, and ever-changing' relations between cytoplasm and nucleus:

> There cannot be any absolute separation between the functions of the nucleus on the one hand and the functions of the cytoplasm on the other. Their relations are reciprocal, each affecting each in constant succession. Nor can either be understood save in relation to the other, and to the activity of the cell as a whole, for neither is capable of long-continued existence apart from the rest of the cell. To establish then a rigid distinction between the nucleus and the cytoplasm, to allot to each element clearly defined and separate functions, is to deal with unreal abstractions. To regard one as controlling the other is quite illegitimate and introduces that dualism of agent and thing acted upon which runs through and vitiates all theories of nuclear dominance. (Russell 1930: 157)

Similarly, after citing a passage from E.G. Conklin urging the view that the preformed cytoplasm directs the egg and sperm nuclei, Russell underscores that 'it is the entire cell, both nucleus and cytoplasm, that is concerned in heredity and differentiation' (Russell 1930: 87). Furthermore, 'we do not consider for example, like Conklin and Loeb, that the "embryo in the rough" is determined by the cytoplasm only, any more than we agree that the chromosomes are solely responsible for the finer characteristics which appear later in development . . . For us, nucleus and cytoplasm are indissolubly wedded in their action upon development' (Russell 1930: 284). As a result, Russell's is not a particulate account of development or of heredity; it is, rather, systemic – and therefore stands in contradistinction to the modern consensus regarding both nuclear animism and nuclear primacy (the precursors of genetic animism and genetic primacy).

Second, recalling the early twentieth-century redefinition of heredity to exclude development, a dichotomy followed not only between preformation and epigenesis but also between nucleus and cytoplasm (Maienschein 1986). However, for Russell, there was no such dichotomy and thus there should have been no division of labour between geneticists (and evolutionists) and embryologists. Russell's insistence that heredity is a feature of development therefore prescribes an alternative to what eventually became the Modern Synthesis of transmission genetics (population genetics) and evolutionary biology. That said, the conceptual divorce between development and heredity was pragmatically necessary, such that Russell's proposals were unworkable in 1930. Not so today.

On Russell's organismal approach to biology it is impossible to ignore development, for heredity itself is a feature of the development of the whole organism. For Russell, the remarkably true repetition of species-specific type is the fact of heredity; but it is also the goal (or natural purpose) of the organism: 'the unique character of the living individual as the fundamental unit of biology stands out unmistakably, for the individual is essentially a functional unity, whose activities are co-ordinated and directed towards the development, maintenance, and reproduction of the form and modes of action typical of the species to which it belongs' (Russell 1930: 166). Despite his characterisation of reproduction as 'one of the master-functions of the organism, in a sense the crown and completion of individual development', Russell laments the way in which 'reproduction has ceased to be taken seriously as a primary biological problem, ever since the general acceptance of the germ-plasm theory' – a trend that persists (Russell 1930: 9, but see Griesemer 2000). Reproduction is a whole-organism activity, requiring all the diverse resources of a whole organism for initiation and maintenance and resulting in the production of a whole organism – but currently biology is theoretically ill equipped to deal with whole organisms, trading as it does mainly in genotypes and phenotypes.

This latter claim helps to explain the (sociological, if not logical, ontological, or epistemological) success of the notion of a genetic programme, as against the relative insignificance of a developmental programme. However, given the difficulties with the former notion, it is imperative to marshal a case in favour of something like the developmental programme alternative.

Keller (2001) takes a crucial first step in this direction with her 'beyond the gene but beneath the skin' approach to biology, which I will explicate and elaborate in Chapter 5 as part of a synthetic theory of creative development. First, however, I must elucidate my notion of constitutive epigenetics which underwrites that account and serves to finally displace genetic primacy.

CONSTITUTING GENES

It is noteworthy that a set of orchestral metaphors has recently been employed in order to move away from earlier notions of genetic control, and of genes as programmes, blueprints, or instructions. For instance, Mayr has remarked that 'by necessity, the analysis of genes and gene-controlled biochemical processes had to be reductionist at the beginning, but it was soon realized that the genes interact with one another and with the cellular environment, much like musicians in an orchestra. The study of this well-orchestrated interaction of genes and cells during the making of an individual is currently the frontier of developmental biology' (Mayr 1997: 152–153). Steven Rose has suggested further that

> far from being isolated in the cell nucleus, magisterially issuing orders by which the rest of the cell is commanded, genes, of which the phenotypic expression lies in lengths of DNA distributed along chromosomes, are in constant dynamic exchange with their cellular environment. The gene as a unit determinant of a character remains a convenient Mendelian abstraction, suitable for armchair theorists and computer modellers with digital mind-sets. The gene as an active participant in the cellular orchestra in any individual's lifeline is a very different proposition. (Rose 1997: 125–126)[11]

Jablonka and Lamb (1995) offer a version of this musical metaphor some-where between Mayr and Rose, which is then extended and elaborated by Keller:

> If the score represents hereditary information in DNA, the phenotype is a specific interpretation of this score at a certain time by certain artists. The interpretation does not affect the score. However if there is another transmission system – recordings – through which a particular interpretation can be transmitted from generation to generation along with the written score, the situation is rather different. There can then be evolution of interpretations of the score, based on the influence that one interpretation has on subsequent interpretations, and that these have on still later ones, and so on. Both the phenotype (the present interpretation) and the genotype (the written score) influence subsequent interpretations. (Keller 1999: 114)

For Jablonka and Lamb, the phenotype of the gene – instantiating epigenetic processes such as chromatin marking and genomic imprinting – is this alternative transmission system. But, as Keller notes, 'to do justice to their full argument, their analogy should have been taken further'.

> Not only does the phenotype (the present interpretation) influence subsequent determinations through epigenetic inheritance, but it can also participate in the modification of the genotype (the written score) itself – as if, e.g., marks were inserted in the score in response to current interpretations. For Jablonka and Lamb, the real (and most radical) conceptual payoff comes not so much from the existence of multiple inheritance systems as from the interaction with actual nucleotide sequences. (Keller 1999: 114)

What I take Keller to be emphasising here is that the epigenetic inheritance systems of which Jablonka and Lamb write are not best thought of as 'in addition to' genetic inheritance systems; rather, the two systems are in a non-additive relationship of interaction. I would go still further: there is neither score nor recording except in performance; the orchestra and conductor together create the score anew with each performance. In other words, epigenetics is constitutive, not additive.

According to a constitutive account of epigenetics, epigenetics does not reduce to gene regulation, for genes themselves do not pre-exist developmental processes. The starting point of epigenetic control cannot be the genome, for the genome does not precede the cell-organism, nor is the latter ever coextensive with or delimited by the former. Constitutive epigenetics is therefore not a genes-*plus* account of epigenetics.

My definition of constitutive epigenetics is as follows: epigenetic events are developmental interactions within the whole cell-organism in its developmental context, between any and all of such factors as cytoplasmic structures, DNA sequences, mRNA, histone- and non-histone proteins, enzymes, hormones, positional information, parental effects, temperature cues, and metabolites. Many epigenetic structures are not stable and do not pre-exist the interaction but rather emerge from these interactions in ontogenetic space and time (Burian 1997: 259–260; Oyama 2000b: 84). These interactions *generate genes*, which are not sequences of DNA encoding amino acid sequences[12]; on my account, a 'gene', *qua* developmental unit, is assembled from multiple resources including nucleotide bases scattered throughout the chromosome and used at particular places and times as a template to co-generate functional, folded, three-dimensional structures – and so co-generate linear polypeptide chains along the way (Neumann-Held 1999: 129; Griffiths and Neumann-Held 1999).[13]

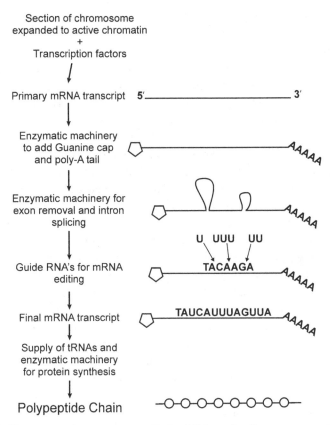

Section of chromosome
expanded to active chromatin
+
Transcription factors

Primary mRNA transcript   **5′**_____ **3′**

Enzymatic machinery
to add Guanine cap
and poly-A tail

Enzymatic machinery for
exon removal and intron
splicing

U UUU   UU

Guide RNA's for mRNA
editing   **TACAAGA**

Final mRNA transcript   **TAUCAUUUAGUUA**

Supply of tRNAs and
enzymatic machinery
for protein synthesis

Polypeptide Chain

Figure 6. Gene expression or gene constitution? Schematic diagram representing the context dependency of gene expression, emphasising transcription, editing, and translation processes. See the text for details regarding why I interpret these processes as gene-constitutive and not merely gene-expression processes. Redrawn with permission and modified from Figure 1 in Griffiths and Neumann-Held (1999: 658).

Both the structure (the particular assembly) and the function (the onto-genetic role) of genes derive from spatial and temporal aspects of the state of the cell-organism. In turn, genes so produced help to regulate ontogenetic processes in the developing organism as participants in nonlinear feedback and feedforward networks generating and being generated by the developing organism. Consequently, the usual idea of genetic primacy is rendered incoherent.[14]

Genes-*plus* approaches to ontogeny illegitimately privilege one factor in a complex network of interacting factors. Developmental control resides neither in the genome nor in the extragenomic epigenetic system. The idea that exclusive control resides within one component of a developing system is in direct conflict with the basic sense that development is in fact *co*-actional. On both sides of the debate, theorists often suggest that either genes or cellular environments are more important; this is especially so in the case of gene centrists, who tend to proceed in their experiments as though the genes were crucially and predominantly important and who tend to relegate the other necessary factors to background or standard conditions. This is an instance, in my view, of not taking development seriously by refusing to grasp the constitutive nature of epigenetics (and the creative nature of development).

In contrast, consider the theoretical elaboration of the developmental systems approach offered by Neumann-Held (1999), elements of which I have incorporated into my account of constitutive epigenetics. She holds that the developmental systems approach (as instantiated, for instance, by Gray 1992 and Oyama 1985) grants too much to genetic primacy theorists in focusing mainly on the *functional* aspect of the classical molecular gene concept and ignoring problems with the *structural* aspect of the concept. 'In modern textbooks of molecular biology a gene is defined as a certain segment (that might be interrupted) of the DNA, which has the function to code for a linear polypeptide chain, regardless of how complex the mechanism of expression might be' (Neumann-Held 1999: 114). Gray adopts this perspective: 'the nucleotide sequence does specify the primary structure of a protein', thereby assuming a (more or less) simple correspondence between a DNA segment and a linear polypeptide chain (Gray 1992: 170). So, too, does Oyama, who notes that 'it makes sense in general to say that the primary structure of a polypeptide is encoded on the chromosomes' (Oyama 1985: 70). But Neumann-Held contends that, given current results in molecular biology, 'it is not necessary to make any concessions to the structural aspect of the gene concept', especially because any such concession may be interpreted as an implicit endorsement that structure determines function – as against the grain of the developmental systems approach (Neumann-Held 1999: 115).

Neumann-Held shows that the very structure of genes is deeply context dependent, caused by such processes as mRNA processing and mRNA editing (see Figure 6).[15] The phenomenon of mRNA processing shows that, in the process of gene expression, DNA is not a unique carrier of developmental information. A simple example will suffice: 'In eukaryotes, the so-called 5′ end of the transcript is "capped" by methylated Guanine, whereas the other end of the transcript, the 3′ end is shortened by a few nucleotides, whereafter a tail of about 200 adenine nucleic acids is added (polyadenylation). These modifications, which are catalyzed by specific enzymes, are not prescribed in the DNA. However, they are essential for further processing (including translation) of the mRNA in the eukaryotic cell' (Neumann-Held 1999: 121).

Meanwhile, though we find it convenient to think that something as basic as the sequence of nucleic acids of DNA is given (hence the Human Genome Project), Neumann-Held underscores that 'as a matter of fact, it is not. Probably the most unbelievable kind of obtaining different polypeptides from the same DNA segment is mRNA editing in mitochondria and chloroplasts', and it has also been 'shown in *Physarum polycephalum* [slime molds], mammals, viruses and higher plants'. The phenomenon of mRNA editing can be divided into two distinct kinds of phenomena. 'In one kind, nucleotides are inserted into or removed from the mRNA. The second kind converts nucleotides, for example, C(ytosine) in[to] U(racile) (and the other way around)... These processes [of mRNA editing] can only be described in the following way: environmental (developmental) conditions, primary mRNA, and processes such as mRNA editing, are in reciprocal ways contingent on each other in the determination of the structures that become translated' (Neumann-Held 1999: 122–124).

On the basis of her discussion of mRNA processing and mRNA editing and other elements of the mechanics of transcription and translation, Neumann-Held concludes that 'regulatory sequences and coding regions do not exist in the DNA or mRNA independently of the system... On the contrary, regulatory elements and coding regions are co-constructed (in a structural and functional sense) in reciprocal contingency by the components of the systems in succeeding processes' (Neumann-Held 1999: 124). In other words, the very structure as well as the ontogenetic meaning of a bit of DNA is constituted by the (spatial, historical, temporal, environmental, and organismal) interactive developmental context in which it finds itself. The genome simply does not precede any (additional) element of the developing organism, nor is it ever identical with the organism. The organism, therefore, as well as the evolutionary history of its species, precedes the genome, with which it immediately enters a complex array of constitutive epigenetic interactions.

The organism, in context, was there all along. Hence Russell's prescient observation, already cited: 'the unity of the organism is not something which comes to be during the course of development, but is there *ab initio*'. That is the only primacy thesis worth defending.

5

Creative Development

What comes of the chemical, mechanical, and social-psychological resources an organism inherits depends on the organism and its relations
with the rest of the world. It makes its own present and prepares its
future, never out of whole cloth, always with the means at hand, but
often with the possibility of putting them together in novel ways.

– Susan Oyama (2000a)

In this short chapter, I draw on the criticisms of the modern consensus offered
in Chapters 3 and 4, and so on my account of constitutive epigenetics, to produce a framework for understanding and explaining organismal development.
Within this framework, genes play an important role, but as derived rather
than driving factors; here, developmental agency is restricted to organisms.

I begin with a brief discussion of the benefits and limitations of developmental biology based on model systems, underscoring the lesson of Chapter
1 that we must be careful in making scientific generalisations on the basis
of particular (types of) experiments. I then emphasise again the notion that
the organism is the basic unit of development, and I proceed to elaborate my
framework for understanding and explaining development in creative terms. In
showing how and where this framework differs from the modern consensus, I
discuss the dialectics of gene–organism–environment interactions in development, with particular attention to the phenomenon of niche construction. I then
emphasise the necessity of taking a systems perspective on development, and I
explore a recent account of how to model these interactions in a systems mode.

THE ORGANISM IN CONTEXT

As I urged in Chapter 4, the cell-organism is the primary unit of development.
Yet, the determinants of development – of the structure and function of genes,

for instance – need not always be found exclusively within the integument of the cell or the skin of the body. In his paper, 'Ecological Developmental Biology: Developmental Biology Meets the Real World', Scott Gilbert (2001) provides an account of how to integrate a wide range of causal factors in development. The subtitle draws attention to the fact that as developmental biology is usually studied, it is insensitive to environmental factors in ontogeny. Such insensitivity is, of course, intentional – it is a means to experimental tractability, evidenced most clearly in developmental biology's predominant focus on a half-dozen inbred model species: the nematode worm, *Caenorhabditis elegans*, the fruit fly, *Drosophila melanogaster*, the zebrafish *Danio rerio*, the chick *Gallus gallus*, the house mouse, *Mus musculis*, and the frog, *Xenopus laevis*. Of course, developmental biologists study many other organisms as well, such as squid, tunicates, horseshoe crabs, and dogfish. However, the six model species just noted are those used most widely and are so popular for a number of reasons, including their small body size, fecundity, early sexual maturation, and developmental laboratory friendliness. As model systems, they have been 'selected for their suitability to the genetic paradigm of developmental biology' and are particularly insensitive to the sorts of environmental wrinkles that tend to muck up the process (Gilbert 2001: 3; also see Bolker 1995; Bolker and Raff 1997; and Gilbert and Jorgensen 1998). As indicated in Chapter 1, the sorts of claims sometimes made about the roles of genes in development may be possible (in principle, anyway) when made about gene action in such highly derived model systems against a constant background of non-specific enabling conditions; but those claims are impossible in less contrived circumstances, such as those obtaining in the real world.

Gilbert explores what developmental biology might look like should it engage ecology in a sustained manner. Two key areas of concern include the study of developmental plasticity and the environmental context dependency of development. To be sure, standard developmental biology may well be attentive to these areas (Hall 1999); but ecological developmental biology (or 'eco–devo') heightens their importance in several ways.

Developmental plasticity refers to the simple fact that there is no one-to-one relationship between a particular genome and a particular phenotype; a single genome may be associated with any number of phenotypic variants, such that the phenotypic expression of a genome is the product of a system of epigenetic interactants coming together over a life cycle. Moreover, not only is the genome–phenotype relationship one to many, it is also many to one, as the same environmental conditions may generate the same phenotypes from different genomes. When a single genome is studied across a range of environmental conditions, the resultant phenotypes will often be quite different from one another; such studies may be represented graphically in the form of a norm of reaction which precludes claims that a genome produces a particular phenotype unless the local environmental conditions are also specified (Schlichting and Pigliucci 1998; Sarkar 1999; Pigliucci 2001).

A nice instance of this type of developmental plasticity is the phenomenon of polyphenism, which refers to the occurrence of either/or phenotypes from a single genotype in a single population. Two morphs, produced from the same genotype, may be so substantially phenotypically different that they may be mistakenly thought to be of different species. Consider the European map butterfly, *Araschnia levana*, which affords two seasonal phenotypes. The summer morph is primarily black with a white band pattern, whereas the spring morph is orange spotted with black. The developmental difference between the two phenotypes is generated by ecological conditions – day length and temperature in the larval stage – and by manipulating the ecological conditions in the lab, researchers can make summertime caterpillars give rise to springtime butterflies (Nijhout 1991). Seasonal polyphenism is but one type of polyphenism; other types include polyphenisms induced by population density, social structure, nutritional factors, and the presence of a predator.

Population Density

Consider that in Mermethids (a family of roundworms that are parasitic on insects in their larval stage but free living as adults), the number of worms developing on a host, and hence the amount of host available to each worm, determines the sex of the worms: if only one to five, then almost all of them are female; if greater than sixteen, then almost all of them are male; in between five and sixteen worms, the proportion of males increases with the number of parasitic larval worms (van der Weele 1999: 106–107).

Social Structure

In the fire ant, *Solenopsis invicta*, there are two forms of social organisation – monogynous colonies with a single reproductive queen, and polygynous colonies with many reproductive queens. The queens of monogynous colonies are heavier and have larger fat reserves than their polygynous conspecifics, but these physiological and morphological variations (which arise from the same genotype) are induced by the type of colony in which individual queens happen to be raised. So social organisation (probably through the intermediary of pheromone exposure) induces morphologies and physiologies which are themselves determinative of social organisation in new colonies (Keller and Ross 1993; also see Gray 2001).

Nutritional Factors

In the moth *Nemoria arizona*, for instance, seasonal diet determines the colour of the moth thanks to tannin content. Spring caterpillars eat oak catkins and develop into catkin-coloured moths; summer caterpillars eat oak leaves and develop into oak-twig-coloured moths (van der Weele 1999: 107–108).

Predator-Induced Polyphenism

The morphology of *Daphnia*, the water flea, will be altered if the fleas develop in water in which their predators have been reared. If juvenile *Daphnia* are made to develop in water in which the predatory larvae of *Chaoborus* (a dipteran) have been cultured, the presence of chemicals released into the water by the *Chaoborus* may induce development of a helmet during *Daphnia* development (see Figure 7). The helmet permits easier escape from the predator, and so the morph benefits from the ecological induction; but the

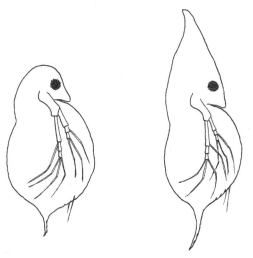

Figure 7. The water flea, *Daphnia cucullata*. Left: *Daphnia* morph not exposed to predators or predator-linked chemicals. Right: helmet-bearing *Daphnia* morph reared in kairomone-laced water. Kairomones are released by the predatory larvae of *Chaoborus* which preys on *Daphnia*; the kairomone-induced helmets (or neck spines) facilitate predator avoidance. Redrawn and substantially modified from the photographs in Figures 1c and 1d in Gilbert (2001: 2).

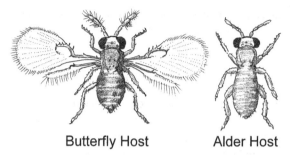

Butterfly Host   **Alder Host**

Figure 8. The parasitic wasp, *Trichogramma semblidis*. Left: Wasp reared in a butterfly host. Right: Wasp reared in an alder fly host. Redrawn with substantial modification from Figure 12-6 in Gottlieb (1992: 153).

production of helmets may also limit resources for provisioning eggs, which creates a developmental–evolutionary trade-off (Gilbert 2001: 4).

Such instances of polyphenism, and so of developmental plasticity, are also instances of the more general phenomenon of the ecological context dependency of development. As the phenotype is neither explicable nor predictable from the genotype alone, its character is dependent upon the context of development (which includes, but does not reduce to, the genome). A striking example of context dependency is the parasitic wasp *Trichogramma semblidis*, whose eggs are lain sometimes in a butterfly host and sometimes in an alder fly host (see Figure 8). Drastic phenotypic differences result from the developmental context (in this case the host), so drastic, in fact, that the adult wasps were once thought to be of two different species (Gottlieb 1992: 153).

Of course, not only is normal development affected by non-genetic causal factors; so too may the environment generate developmental abnormalities. As biologists explore development in new non-model systems, and also complicate their experiments with the model organisms already well studied though context independently, new research programmes and, correlatively, previously unanticipated findings, begin to emerge. As a final example, consider that recent research has indicated that several pesticides that appear to be relatively safe when amphibians are exposed to them under laboratory conditions in fact turn out to be potentially devastating to the same creatures when exposed in the wild. Gilbert (2001: 6) notes that the compound methoprene functions as a juvenile hormone mimic in mosquito pupae, preventing their metamorphosis into adults. As vertebrates do not contain juvenile hormone, a standard assumption was that this pesticide simply would not be harmful to fish, reptiles, or humans. This assumption holds up in the lab: methoprene, itself, is not a teratogen – that is, methoprene, as such, does not cause developmental malformations. However, upon exposure to natural sunlight, methoprene decomposes into two substances that generate substantial developmental abnormalities in *Xenopus* tadpoles.

Biologists from many disciplines have begun to show that such hormone-disrupting effects are associated with pesticides and other industrial chemicals in widespread use (Krimsky 2000; the discussion in the next four paragraphs closely follows Robert 2001b). The environmental endocrine hypothesis asserts that a wide range of chemicals have the capacity to imitate or antagonise normal hormone function, leading to observable developmental and reproductive effects in animals, including humans. The endocrine system hosts a complex dance of hormones, such as estrogen, testosterone, progesterone, and follicle-stimulating hormone, which work by binding to specific receptors either inside or on the surface of cells. Once the hormone attaches to its receptor, a variety of biochemical effects follow, all of which are crucial to the generation, development, sustenance, and eventual reproduction of an organism. The endocrine disruptors, as they are commonly known, are able to obstruct the usual flow of hormonal information and even to mimic hormone activity, though not perfectly, thereby altering the basic biological role of the endocrine system.

Hormone disruptors function through a variety of mechanisms and with a range of effects, depending on which toxin is present, which particular hormone is being mimicked or obstructed, the level of exposure, and the time of exposure (embryos exposed to a tiny amount of dioxin during a particular developmental window will be affected very differently from adults exposed to the same or a larger amount). Some of the documented effects of endocrine disruptors in wildlife and animal models include the following: abnormal testicular and ovarian development; testicular, prostate, breast, and ovarian cancer; feminisation or demasculinisation; endometriosis; and reductions in sperm counts. A number of researchers have hypothesised similar effects in humans, but there is as yet little incontrovertible evidence that humans have suffered from hormone disruption.

The scientific evidence for humans is as yet somewhat shaky for three main reasons: the ethical unacceptability of the sorts of studies that would be required to definitely establish a causal relationship in humans; the relatively long lag time between exposure to a particular toxin and its possibly causally related effect; and endocrine disruptors' defiance of the basic framework of toxicology. The latter problem is indeed tractable, but the fact that endocrine disruptors challenge the standard monotonic dose-response curve (increased doses lead to larger effects) helped to delay the formulation of the environmental endocrine hypothesis. Hormonally active agents play a role in destabilising otherwise self-regulating feedback and feedforward systems, and tiny doses may have large effects whereas large doses may have smaller effects. Moreover, the time at which the dose is administered is crucial in mediating its effects, as is the overall state of the system. Well-designed studies within a broader framework of causal systems should help to alleviate some of the scientific instability of the environmental endocrine hypothesis, although extrapolating from wildlife studies and experiments on animal models to humans – given the impossibility of performing the rights sorts of experiments on humans – opens the door to scientific concern (and also political scepticism masquerading as scientific concern). Thus, Gilbert's call for greater interplay between developmental biology and ecology is both timely and welcome.

This emphasis on environmental causes in ontogeny reflects attention to factors well above the level of the genome (van der Weele 1999). Of course, this is not to say that the genome itself is not crucially implicated in the development and reproduction of organisms, but only that the genome does not exhaust the corral of causal complexities. Nevertheless, some biologists tend to operate as if the latter were in fact the case. Witness the trend in developmental biology toward the study of the actualisation of genomic potential. Maynard Smith, although suggesting that there is 'a lot more to development than gene regulation', contends in the very next breath that, 'in particular, there is the question of how genes get switched on in the right places' (2000b: 218). However, as Gilbert's recent review attests, classes of non-genetic developmental causes that both interact with and operate relatively independently of the genome have been long recognised. This is easy to forget, inasmuch as organisms are too often to this day conceived as the product of genes – acting and being activated in environmental contexts, of course, though the genes are typically seen as of primary determinative importance.

In this regard, Evelyn Fox Keller (2001) has written forcefully about the elision of the organismic body in modern biology. The body, when seen at all, is seen passively, as a nurturing environment for the active, ontogenetically and evolutionarily important work of the genes. To be sure, when we transcend the nature–nurture or gene–environment dichotomy in favour of a conjunction of the two, then we prima facie develop a concern for both genes and environments; that is the force of the interactionist consensus. But is the body best construed as an *environment* for genetic activity? Or rather should the body, in its various environmental contexts, be conceived more actively (and interactively) as a developmental agent?

Keller (2001) notes an ambiguity over identifying 'the body', despite its crucial evolutionary and ontogenetic significance. It is something within an outer integument, of course, but is the body the multicellular organismic body contained by the epidermis, the cellular body contained by the cell membrane, or the nuclear body contained by the nuclear membrane (in eukaryotes)? Keller chooses to focus on the cellular body at that time in the life cycle when it is coextensive with the organismic body – the zygote or fertilised egg, porous and permeable as it may be. Like E.S. Russell, she holds that 'the true germ-plasm must be the cell-organism' (Russell 1930: 193). One reason is that the cellular integument serves the vital function of holding things together in close proximity:

> Proximity is crucial for it enables a degree of interconnectivity and interactive parallelism that would otherwise not be possible, but that is required for what I take to be the fundamental feature of the kind of developmental system we find in a fertilized egg, namely, its robustness. Prior to all its other remarkable properties – in fact, a precondition of these – is the capacity of a developmentally competent zygote to maintain its functional specificity in the face of all the vicissitudes it inevitably encounters. (Keller 2001: 301)

Keller's ideas here are consilient with Russell's notion of the autonomy of the developing organism, as explored in Chapter 4. The autonomy of the organism refers to its ability to reach a developed form despite, as it were, environmental vicissitudes (Russell 1930: 6–7). However, 'robustness' is a better characterisation than autonomy, considering (a) the deep interrelations – supportive, constitutive, and constraining interactions – between a developing organism and the diverse elements of its various developmental contexts; and (b) the existence of multiple developmental and regulatory pathways toward equifinal endpoints. Accordingly, Keller posits the robust cell-organism as the unit of development, as against either the gene or the genome.

Regarding (a), developmental interactions might be construed in terms of *autokoenomy*, Sarah Hoagland's term denoting 'a self who is both separate and related, a self which is neither autonomous nor dissolved: a self in community who is one among many' (Hoagland 1988: 12). I borrow the concept in this context to capture the nature of the phenomena of morphogenesis, differentiation, and growth. Consider the cell in development: though an individual, the cell is also at base an interactor; the cell's ontogenetic progress is deeply bound up in interactions with the activities of other cells and developmental factors and entities. Inasmuch as autokoenomy is used by Hoagland to describe selfhood and conscious, intentional action, I run the risk of anthropomorphising developmental entities and processes in describing them as autokoenomous. It's a risk worth taking, though – and one not uncommon in biological literature (*viz.* the depiction of aspects of development as self-organising).

Regarding (b), developmental redundancy is absolutely crucial (Strohman 1993; Gilbert et al. 1996). Gabriel Dover (2000), for instance, defines robustness as 'the ability of a system to continue functioning despite substantial changes to its components' through various processes such as internal coevolution (Dover 2000: 1157, 1158). Relatedly, Wallace Arthur (2000: 51) notes that 'genes are changed by mutation. Populations are changed by selection (and drift). Development is changed by – what?' In response to this question, Arthur underscores the process of 'developmental reprogramming' – I prefer developmental reorganisation – which fills the gap between mutation and selection (and drift) by permitting the creative generation of new interactive pathways without sacrificing functionality. Developmental reorganisation functions here as a 'naval engineer' on Theseus' ship by changing the design and construction of the organism while it develops, without sacrificing its survivability (also see Budd 1999; Dover 2000: 1158).

CREATIVITY IN DEVELOPMENT

Recall Keller's distinction between genetic programmes and developmental programmes, discussed in Chapter 4. Within the cell-organism, the genome is of vital significance to a developmental programme dispersed throughout the whole organism, but the genome does not itself contain or comprise a programme for development. Keller argues convincingly that developmental information is not encoded in the genes but is instead distributed throughout the fertilised egg undergoing ontogenesis. The developmental programme is not composed of particular genetic entities, and it does not reside in the genome itself; rather, it consists and exists in 'the cellular machinery integrated into a dynamic whole'. Thus, Keller writes that 'if we wish to preserve the computer metaphor, it would seem more reasonable to describe the fertilized egg as a massively parallel processor in which "programs" (or networks) are distributed throughout the cell' processing nuclear and cytoplasmic and other bits of developmental data (Keller 2001: 302, 307).[1] The cell-organism is therefore both ontogenetic agent and a material source of developmental information. As against the modern consensus, then, the cell-organism is a contextualised generative entity conditioning, and only partly conditioned by, its genes. Accordingly, development is not strictly or primarily a genetic process but rather a function of the whole organism.

Hence, the creativity of development: drawing on the contextually conditioned nuclear and cytoplasmic structure it inherits along with much else,

86

a developing organism constructs, processes, and regulates specific ontogenetic resources dispersed throughout itself and its environment. Development, to use Russell's phrase, is a 'living, responsive activity of the organism' (Russell 1930: 109). In this sense, then, organismic development is an autopoietic[2] (self-constructive) process not only post-natally (when it is obvious that organisms creatively construct themselves), but also from conception (Gottlieb 1971).

Development will be taken with due seriousness only when it is consistently acknowledged, in theory and in practice, that organisms are not the product of epigenetically triggered, preformed genetic programmes. For ontogenesis as a creative process is something more, in fact something other, than differential gene expression, however chaotic, non-linear, or emergently epigenetic. Given the account of constitutive epigenetics elaborated in Chapter 4, we see that genetics reduces to epigenetics (development) and not the other way around (Griesemer 2000). Development is not a matter of genes-*plus* anything but rather a matter of the organism's semi-autonomous self-constitution from a range of ontogenetic raw materials.

It is evident not only that organisms construct themselves within environments but also that they help to construct their environments. Both processes are creative, in the sense that something new emerges from them: a new organism in a new environment, each contributing to the construction of the other in a synergistic, coevolving dyad. Note first that organisms inherit much more than genes at conception, birth, and beyond: we inherit, for instance, complex cellular structures and a structured embryonic and foetal stimulative environment (including metabolites and other nutrients, temperature cues, and behavioural stimulants). But we inherit still more: parents, conspecifics, and their various habitats – West and King (1987) use the phrase 'ontogenetic niche' to specify inherited species-typical legacies of society and ecology. And yet although this ontogenetic niche contributes significantly to individual development, it is not quite ready-made; it is rather always in process, acted upon by organisms of all stripes, and mutually acting upon them.

Here we see Richard Lewontin's signature emphasis on gene–organism–environment dialectics (e.g., Lewontin 1983). Lewontin was objecting to the latent externalism of evolutionary theory, namely the presumption that evolution works by a problem–solution mechanism: the environment poses some problem for organisms, and organisms, through trial and error, provide a solution to that problem, leading to a more adaptive phenotype. Lewontin proposed that organisms do not adapt to external environments so much as construct those environments in a reciprocal, dialectical interplay. Several biologists have undertaken to model such dialectical relationships in an effort both to render these ideas more concrete and also to better explain the dynamics of evolutionary processes in the context of genetics (and development).

The basic impulse behind models of niche construction is that organisms, through their activities and through their metabolism, define their effective environments in both creative and destructive ways (Odling-Smee et al. 1996; Laland et al. 1999, 2001). Kevin Laland, John Odling-Smee, and Marcus Feldman define niche construction as occurring when 'an organism modifies the functional relationship between itself and its environment by actively changing one or more of the factors in its environment, either by physically perturbing these factors at its current address or by relocating to a different address, thereby exposing itself to different factors' (Laland et al. 2001: 118). Darwin (1881) offered his own example of niche construction, describing the burrowing activities of earthworms which alter the structure of the soil; but earthworms also transport organic material into the earth which mixes with inorganic material, and they stimulate various sorts of microbial activity, thereby altering the biochemistry of soils as well. These structural and biochemical effects accumulate over generations, such that each new generation of earthworms inhabits a new environment and is thus subject to variable selection pressures (Odling-Smee et al. 1996: 641). Many other examples are evident throughout both the animal and plant kingdoms.

Laland, Odling-Smee, and Feldman have built two-locus population genetics models to explore the evolutionary effects of niche-constructing behaviour. In one such model (Laland et al. 1999), they investigated the dynamics produced by niche construction leading to either an increase or a decrease in the availability of an environmental resource (the availability of the resource itself may vary independently of niche construction, and the model builds in this possibility). The model shows that the effects of the niche-constructing behaviours may supersede external selection pressures; for instance, where niche construction generates selective pressures at odds with the action of an external selection pressure at one of the loci, a likely outcome is the fixation of otherwise deleterious alleles in a population (Laland et al. 1999: 10, 246). Furthermore, as the new selection pressures persist transgenerationally, unusual evolutionary dynamics are possible. For instance, time lags were observed between the emergence of a new niche-constructing behaviour and a population's response to the selection pressures it modifies. These time lags may, over time, generate either evolutionary inertia or evolutionary momentum: in the former case, increasingly strong selection is required to dislodge a population from equilibrium; in the latter case, populations continue to evolve along a particular trajectory even when selective pressures change or are reversed.[3] The evidence emerging from models of niche construction points to the conclusion that 'adaptation ceases to be a one-way process, exclusively a response to environmentally imposed problems, and instead becomes a two-way process, with populations of organisms setting as well as solving problems' (Laland et al. 2001: 122).

It is evident that, thus far, the primary emphasis of those building such models is on the evolutionary consequences of niche construction: the existence of many sources of natural selection which help to shape organisms depends largely on the niche-constructing activities of those organisms and also of their forebears. Here is evidence, therefore, of multigenerational nongenetic inheritance affecting both the development of organisms and their subsequent evolution. Moreover, niche-constructing organisms cannot simply be regarded as vehicles for genes – as in Richard Dawkins' (1976: 21) famous image – for they must be considered responsible for the changes they have wrought in their environments (and in the sources of selection operating therein) which may well be translated into evolutionarily significant modifications.[4]

But niche construction is ontogenetically important, as well, inasmuch as organisms find, make, and provide for themselves and each other some of the environmental resources necessary for their successful development and reproduction, and also may alter the conditions under which they evolve. Again, we witness the creativity of the developing organism at work. Although some of the resources for development are just there, waiting to be used or acted upon, the bulk of them are rather basic products of organismal reproduction. This is not to suggest that genes are not important developmental interactants but rather that genes simply cannot be foundational. The genetic component of ontogenesis is constituted epigenetically during the development of the organism.

Accordingly, if ontogenesis is an additive process, then it is an *organism-plus* process – the modern consensus has things the wrong way around. But just as genes cannot exist independently of organisms, so too can organisms not exist independently of genes; the privileging of either one over the other is therefore inappropriate, given their reciprocal contingency. The full range of developmental material is required, and its ontogenetic specificity is negotiated through spatiotemporally sensitive, contingent, constitutive interactions within the organism itself. The result is the emergence of a developed organism bearing a remarkable resemblance to its parents – without recourse to anything like species-specific genetic programmes.

Seen in this light, it is plainly evident that how organisms develop is not predetermined in scope but only in kind, and the kind that is reproduced is not predetermined genetically but rather co-determined generatively. Organisms then creatively construct their own fates.

Where, then, are we now? With the accounts of constitutive epigenetics and creative development offered here and in Chapter 4, and with the developmental considerations explored in Chapters 1 and 3, we have moved beyond the modern consensus. To reiterate: *contra* genetic informationism, genes do not contain all of the relevant specific information required for development, nor are genes uniquely informational; *contra* genetic animism, genes do not contain a programme for development, and if there is such a thing as a programme for development, it is dispersed throughout the developing system; and *contra* genetic primacy, although genes may be *primus intra pares* methodologically, ontogenetically the developing organism (including but not reducible to its genes) has pride of place.

And yet an obvious objection is waiting in the wings: verbal gymnastics of the sort evident throughout these pages are utterly useless to practicing biologists. This is, admittedly, a persistent worry for me. What I have tried to show so far is at least that the charge of semantic sleight of hand can be turned in the other direction. Metaphors of genetic programmes and genetic instructions do not take us very far in understanding or explaining development. At most, they leave the overly optimistic impression that the problem of development is solved, but only by 'side-stepping the task of *developmental analysis*' (Lickliter 2000: 324). Where development as such is the primary *explanandum*, developmental analysis must be the method of choice. This is what it means to take development seriously.

Though my arguments in this book are largely based upon scientific and not philosophical literature, the arguments themselves are philosophical. They are meant to be biologically useful, but they may not be. Allow me to begin to show, then, how they may be of use – a project I take up again in the following chapters.[5]

Johnston and Edwards (2002) have recently published a series of increasingly specific (or 'unpacked') representations of a model of the development of behaviour, one that is generalisable to development as such and that is also consilient with much of what I have said so far. Their model is not meant to specify every molecular or cellular aspect of the complex interactions comprising development; rather it is designed to provide 'a useful intermediate level of detail that captures that complexity while at the same time rendering it reasonably comprehensible' (p. 31). For these authors, genes are not to be considered 'carriers of information or repositories of plans and blueprints' (p. 27) somehow both separate from and yet also directing development, but instead as molecules that are biologically active within the developmental system and have only indirect reciprocal effects via mRNA synthesis (2002: 26, 28; also see Lickliter 2000). Experience, too, has indirect and reciprocal effects on the development of behaviour, mediated through multiple levels of

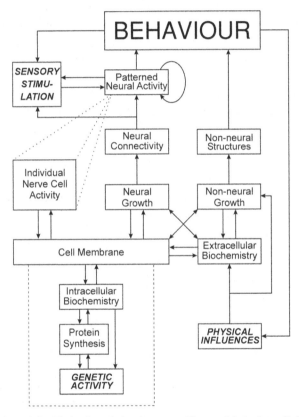

Figure 9. A model of behavioural development. The model depicts all factors, both neural and non-neural, that interact in the developmental production of organismal behaviour. Solid lines with arrowheads between factors represent causal relationships. Dotted lines represent non-causal relationships between patterned neural activity and the activity of individual nerve cells, indicating the nesting of the latter within the former. The elliptical arrow near the top of the diagram depicts spontaneous neural activity. Redrawn with permission from Figure 3 in Johnston and Edwards (2002: 28).

biological, ecological, and social organisation. The model is meant to focus investigative attention on developmental interactions and specific mechanisms, beyond metaphor and shorthand formulations.

Johnston and Edwards' 'completely unpacked model' of behavioural development (Figure 3 in Johnston and Edwards 2002: 28, reproduced here as Figure 9) comprises fourteen boxes, each representing an interacting factor, linked together by means of their various bidirectional interactions (some but not all of which are causal). Any particular instantiation of the model would be only a time slice of a specific developmental moment; the model could be transformed from two dimensions to three with the addition of information regarding the timing of individual influences on development, though this would obviously make it considerably less amenable to pictorial representation (but see Figure 4 in Johnston and Edwards 2002: 29).

The model proposed by Johnston and Edwards can be used to organise existing knowledge and also to make predictions about behavioural development that can be empirically investigated; for instance, the functions (causal or otherwise) represented by arrows or dotted lines connecting factors within the model might capture our knowledge of some developmental process (say, induction) or might 'imply the existence of interactions that would, if they occurred, generate the observed changes' (Johnston and Edwards 2002: 30) – and so produce a new research programme to discover those interactions or, if unsuccessful, lead to alterations in the underlying model. The model is thus a framework for synthesising what we already know about development, but it can also be instantiated under experimental conditions to learn more. Insofar as the account of development offered in these pages can be assimilated into a model such as that by Johnston and Edwards, the account can be put to work by practicing biologists.

However, even were my account of development merely verbal, it nonetheless has a theoretical contribution to make, both in terms of conceiving the aetiology of disease (Robert 2000a, in preparation a) and also in grappling with recent efforts to integrate developmental and evolutionary explanations. In Chapter 6, I introduce the general nature of the 'integrative project' (Sarkar and Robert 2003), explore what I take to be its core premises, and provide several examples to show how focusing on developmental mechanisms is necessary to explaining aspects of organismal evolution. Then, in Chapter 7, I bring the considerations of the first six chapters to bear in empirically and conceptually assessing theoretical frameworks for investigating the relationship(s) between evolution and development.

6

A New Synthesis?

It may seem mystical to suggest the biology is not 'molecular' at its core the way physics and chemistry are. But suppose it is *not* the genome that is especially conserved by evolution. Suppose the ephemeral phenotype really *is* what we need to understand and what persists over time. Genes would then be 'only' the meandering spoor left by the process of evolution by phenotype. Perhaps we have hidden behind the Modern Synthesis, and the idea that all the action is in gene frequencies, for too long.

– Kenneth Weiss and Stephanie Fullerton (2000)

There is much more to both evolution and development than we can learn from focusing primarily on genes.[1] However, this realisation is hard won, given the recent history of biology, and of philosophy of biology. I noted in Chapter 4 that the twentieth century witnessed the biological reconceptualisation of evolution in terms of changes in gene frequencies in a population, and of development in terms of gene expression. Moreover, although philosophers have long been preoccupied with evolutionary theory, and more recently with molecular biology, they have engaged far less frequently with development. Whereas historians of biology have been long intrigued by embryology, philosophers have tended to shy away. Yet times have changed. It is mainly as a result of recent work in developmental and molecular biology that some of the reductionistic biases of genetics have paradoxically come to be seen as constricting future research and precluding genuine understanding of both development and evolution.[2] Moreover, advances in biology generally have permitted us to open the black box of development and to move beyond simplistic models of gene action. (Even with development in a black box, genes alone do not explain development; nor do genes alone explain evolution – see, e.g., Laubichler and Wagner 2001.) As we have come to learn these lessons, a new synthetic framework has emerged within biology, opening logical (and

93

speculative) space for philosophical investigations of the nature of development and its relation to evolution. Thus far, this book has been concerned with development. We now turn to evolution.

Before the rediscovery of Mendel's work in 1900 – and so before the eras of classical genetics, the Modern Synthesis, and molecular and developmental genetics – an *un*modern synthesis flourished in biology, between evolution and embryology (and morphology). The germ layers (now referred to as endoderm, mesoderm, and ectoderm – the layers of cells from which all tissues and organs are formed) were identified and analysed prior to Darwin's theory of evolution (Darwin 1859), but they became absolutely central to biology in the latter half of the nineteenth century as a variety of scientists sought to establish relationships of ancestry and phylogeny through comparative analysis of germ-layer homology between embryos of different species.

An early figure was the German zoologist Ernst Haeckel (1834–1919), who attempted to synthesise Darwinian evolution with comparative morphology and comparative embryology in theorising that 'all multicellular organisms arose phylogenetically from an organism structurally equivalent to the early gastrula, an embryonic stage found early in the development of all multicellular animals'. Haeckel held that all multicellular organisms pass through a two-germ-layer stage, which, he held, was equivalent across species based both on its structure and on the way in which it is produced – the endodermal layer formed from the ectodermal layer by invagination in proceeding from blastula to gastrula. Consequently, Haeckel advocated a theory of ontogeny as the rapid recapitulation of phylogeny in early development, followed by the terminal addition of novelties in the generation of adults (commonly known as the Biogenetic Law). So, according to Haeckel, the ancestral stages of adults could be identified in the embryos of descendants (Hall 1999: 80). Haeckel's theory of recapitulation eventually gave way to less hypothetical accounts of the relationship between embryology and evolution, though the study of germ layers and homology remained essential.

One problem was that Haeckel's theory, like that of Karl Ernst von Baer (1792–1876), required that early embryonic development not be subject to change: hence, Haeckel's account of change based on *terminal* addition. Francis (Frank) Maitland Balfour (1851–1882), for one, saw no reason that early embryonic development should be immutable, and he urged that natural selection could operate at any stage of development from larva to adult (except beyond reproductive capacity): 'I see no reason for doubting that the embryo in the earliest periods of development is as subject to the laws of natural selection as is the animal at any other period. Indeed, there appear to me grounds for thinking that it is more so'. 'The principles which govern the perpetuation of variations which occur in either the larval or the foetal state are the same as those for the adult state. Variations favorable to the survival of the species are equally likely to be perpetuated, at whatever period of life they occur, prior to the loss of the reproductive powers'.[3] Were natural selection to operate early in embryonic development, this would be a function of some larval or foetal variant's capacity to benefit the organism; that is, a pre-terminal variant might be selected and persist, *contra* both Haeckel and von Baer. Walter Garstang (1868–1949) pushed still further, arguing that larval evolution and adult evolution could occur independently. Directly against Haeckel, Garstang urged that 'ontogeny does not recapitulate phylogeny, it creates it' (Garstang 1922: 81).

However, a variety of events conspired against evolutionary embryology at the end of the nineteenth and the beginning of the twentieth centuries. Evolutionists focused attention on the transmission of 'factors' between generations, and they held to an account of development that both took development for granted and also attributed the generation of phenotypic traits to Mendelian genes. Embryologists meanwhile turned to an experimental approach, spurred on by Roux and his programme of *Entwicklungsmechanik*. The conventional story is that, despite persistent efforts on the part of those such as DeBeer, who carried on the project of evolutionary embryology, and despite those such as Conklin, Whitman, and Lillie, who attempted to offer alternative accounts of embryonic development, interest in the relationship between embryology and evolution waned, eventually culminating in the establishment of the embryology-free Modern Synthesis (Hamburger 1980). Despite disputes over the veracity of this conventional account (Amundson 2003), disputes that I am not prepared to adjudicate in these pages, it remains worth asking whether the phenomena of development and evolution are in fact adequately accounted for in terms of the transmission and activation of genes. Given the recent explosion of interest in exploring relationships between development and evolution above the level of genes, and despite the persistence of detractors, the answer would appear to be a resounding No.

In this chapter, I discuss various efforts to integrate evolutionary and developmental explanations. My particular focus is on the project commonly referred to by the sobriquet evo–devo (evolutionary developmental biology);

spanning evolutionary, developmental, molecular, cell, and organismal biology, in addition to genetics, paleontology, and ecology, evo–devo is a new biological synthesis.

As already noted, in the nineteenth and the first half of the twentieth century, connections between evolution and development were established in multifarious productive ways (see, e.g., Allen 1978; Gilbert 1988, 1991a, 1994; Maienschein 1991a; Gilbert et al. 1996; and Hall 1999, 2000a). Unfortunately with the rise of population genetics as our evolutionary paradigm, these connections were lost or forgotten and have only recently been rediscovered and reinvested with scientific importance (Raff 1996; Arthur 1997, 2002; Hall 1999). Like any biological discipline, evo–devo commands a diverse range of theoretical perspectives and experimental approaches. For instance, some evo–devoists focus more heavily on developmental genetics – say, on the roles of homeobox genes in development and evolution (as in Patel 1994 and Carroll et al. 2001), – some less so (witness Gerhart and Kirschner 1997 and Hall 1999, who focus more closely on cells and their interactions). Sometimes, this leads to discrepancies over the nature of the synthesis of evolution and development. It is to these discrepancies that we now turn.

A caveat: different commentators and practitioners describe the relationship between developmental and evolutionary biology in different ways: there is talk of the reconciliation of developmental and evolutionary biology; their integration (or reintegration); the accommodation of one within the other; or their synthesis. In what follows, I will generally prefer the latter locution, synthesis, to describe the discipline of evolutionary developmental biology, though I will sometimes write of the integration of developmental and evolutionary explanations.

A second caveat: evolutionary developmental biology (evo–devo), so named by Hall (1992b), is only one of several projects in this domain; another is developmental evolution (devo–evo), so referred to by Wagner (2000, 2001; Wagner et al. 2000). These are overlapping projects, and the distinction between them is difficult to draw, though the latter appears to be more mathematically oriented. I do not aim in this chapter to settle once and for all any disciplinary disputes between evo–devo and devo–evo. For the sake of convenience, I will prefer the former locution, evo–devo, though I do not intend this to be exclusionary. (In fact, the champion of devo–evo, Günter Wagner, figures prominently in my discussion of evolution and development.)[4]

A third caveat: philosophical debates about development and *evolution* seem often to be rather about development (actually, developmental genetics) and the Modern Synthesis. However, evolutionary biologists have come a long way since the Synthesis; aside from those few biologists who deem the Modern Synthesis unassailable, and also aside from hard-line Neo-Darwinians (gene selectionists, mainly), the views of most contemporary evolutionary biologists have evolved significantly away from the Modern Synthesis (especially with the advent of cladistic analysis, the emergence of comparative genomics and molecular phylogenetics, and so on). Prima facie, it appears that synthesising developmental biology and Neo-Darwinism is a much more difficult prospect than synthesising developmental biology and evolutionary theory more broadly construed. At any rate, *defining* what evolutionary theory today actually *is* is no simple task; in what follows, I sketch an account of Neo-Darwinism that is not up to the challenge proffered by attention to development. I presume that it is non-controversial to suggest that evolutionary theory attends at its core to variation, heredity, and differential reproduction (Lewontin 1970; Wimsatt 2001); explanation of evolutionary change by reference to the mechanisms of natural selection and drift; and tracking of evolutionary change by reference to changes in gene frequencies in populations – although some controversy might persist over the exact referents of these concepts or the appropriate scope of their applicability (Sterelny and Griffiths 1999).

Despite differences in approach, evo–devoists tend to hold to a core of theoretical presuppositions, including: (a) the hierarchical nature of development and evolution; (b) the need to focus on developmental processes – interactions – between genotype and developing phenotype; and (c) the belief that analysing developmental processes and mechanisms, and their evolution, improves our understanding of both development and evolution.[5] Studying development in evolutionary context, and evolution in developmental context, increases the explanatory scope of both sciences. I will focus first mainly on evo–devo's account of development.

Regarding (a), the hierarchical nature of development and evolution necessitates the study of emergent properties inexplicable from lower (or higher) hierarchical levels; for instance, cells' collective behaviour during morphogenesis cannot be explained (or predicted) by examining the behaviour of individual cells prior to cell division, differentiation, or (in animals) condensation – let alone by examining DNA sequences (Hall 2000a: 177). This is because the formation of cell condensations is contingent not on genetic directives but rather on the spatiotemporal state of the organism and its component parts at multiple levels (Laubichler and Wagner 2001; Gilbert and

Sarkar 2000 provide additional examples of such emergent phenomena). Many evo–devoists are thus methodological antireductionists, offering the advice that we must engage in multileveled investigation of ontogeny and evolution in order not to miss key features of either, at multiple hierarchical levels.

Given (a), we see that (b) draws attention to the fundamental ontogenetic importance of epigenetics, whether conventional or constitutive. Some evo–devoists hold that precisely identifying types (and tokens) of epigenetic interactions is central to the project of synthesising development and evolution. Especially important is the investigation of epigenetic interactions within and between modules in morphogenesis. A 'module' is a semi-autonomous component of an organism; modules exist at different levels of the biological hierarchy; and evolutionarily significant within-module changes can occur without disrupting the overall integrity of the organism, thereby facilitating the developmental evolution of novel characters. Bolker (2000) helpfully distinguishes between developmental and evolutionary approaches to modularity and shows how these distinct approaches may interact and coalesce in evo–devo (also see Atchley and Hall 1991; Raff 1996; von Dassow and Munro 1999; Sterelny 2000b; Gilbert and Bolker 2001; and Winther 2001. I discuss modules further later in this chapter and in Chapter 7).

Regarding (c), the belief that analysing developmental processes and mechanisms, and their evolution, improves our understanding of both development and evolution adverts to the conviction that evolution *qua* population genetics, in presupposing development rather than investigating it, tends to miss key elements of evolution. This is not simply a charge of incompleteness but also a charge of explanatory inappropriateness. That is, although it is surely true that the roster of evolutionary change in a lineage is in some ways deficient unless it catalogues changes in developmental pathways as well as changes in adult phenotype and gene frequency in a population (the charge of incompleteness[6]), the further claim of evo–devoists is that the best explanation of evolutionary change is not always made exclusively in terms of changes in gene frequency in a population (just as the best explanation of some developmental mechanism is not always made exclusively in terms of changes in gene expression). Wagner (2000), for instance, has employed the concept of 'explanatory force' (Amundson 1989) to indicate how, in some cases (such as the evolution of stable sex ratios), a population genetic explanation captures the relevant phenomena to be explained better than any competing explanation, whereas in other cases other explanations will be more appropriate. I explore these aspects of (c) in further detail in the next section.

DEVELOPMENT AND EVOLUTION

Evolutionary developmental biologists identify at least two fundamental relations between evolution and development: most evolutionary changes are
introduced during ontogeny, in the sense that ontogenetic modifications, and
modifications in developmental processes, produce evolutionary changes;
moreover, developmental mechanisms themselves evolve. Building on the
example of the blind cave fish introduced in Chapter 2, in this section I
provide three further examples to illustrate the complex interplay between
development and evolution.

Butterfly Wing Morphology

Consider first the development and evolution of eyespot patterns on butterfly
wings (see Figure 10 for an example). Eyespots are a relatively recent innovation, and they are important in predator avoidance as they direct attention
away from vital organs.[7]

One recent suggestion within evo–devo is that eyespot development is
induced by the 'eyespot organizer', a small group of cells which cause adjacent
cells to synthesise pigments (Wagner 2000: 96, referring to work carried out
by Keys et al. 1999 with *Precis coenia* and *Drosophila melanogaster*). The
eyespot organiser appropriates specific molecules involved in establishing

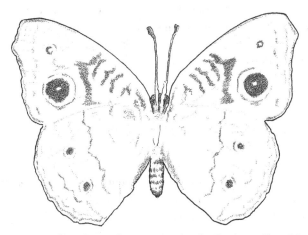

Figure 10. Seasonal linea form of *Precis coenia*, the Buckeye Cape May butterfly.
Redrawn with permission and substantial modification from the photograph in Figure
1a in Brakefield and French (1999: 392).

the basic wing plan of the butterfly – in particular, the anterior–posterior boundary – and attaches to them additional regulatory functions, from which the changes in butterfly wing morphology follow. A population geneticist would explain the emergence of eyespots in terms of a genetic change in the population, tracked by the selection of mutant alleles responsible for the new patterns. But although it may very well be true that one or two mutations are involved, 'to state that a genetic mutation led to a favored character, which, in turn, was selected is utterly uninformative in explaining innovation' – not least because 'the emergence of morphological innovations depends to a large extent on the epigenetic dynamics of the involved developmental pathways' (Wagner et al. 2000: 822–823; see also Pigliucci and Schlichting 1997 and Newman and Müller 2000).

That the population geneticist is mute regarding the developmental biology of an evolutionary change does not make the population geneticist's explanation wrong; it rather evinces that the evo–devo explanation is both more complete and more appropriate (has more explanatory force) in this context. For without detailed knowledge of the developmental interactions between genes and proteins involved in establishing the anterior–posterior boundary in *Drosophila* and butterfly wings, 'it would have been impossible to understand which genetic changes were sufficient to establish an eyespot organizer' – the evolutionary innovation of interest (Wagner 2000: 97). In other words, though the evolutionarily significant change may well be tracked at the genetic level, the change occurs within a developmental mechanism inaccessible (and of little interest) to the population geneticist. Evolutionary innovations, especially in morphology, have been something of a mystery to evolutionary biologists (Mayr 1960) but are more straightforward (though still complex) when examined developmentally.

External Furry Cheek Pouches

Consider next the origin of fur-lined external cheek pouches in geomyoid rodents, mainly pocket gophers and kangaroo rats (Brylski and Hall 1988a, 1988b), as shown in Figure 11. Other rodents have cheek pouches internal to the mouth, which are lined with buccal epithelium; in contrast, geomyoid rodents have cheek pouches opening outside the mouth, which are lined with fur. Both types of cheek pouches are used to store food obtained during foraging, though external pouches may be both larger and more efficient at conserving body water than internal pouches. Drawing on developmental data, Brylski and Hall have shown that internal pouches are the ancestral condition; that is, the evolutionary ancestors of living geomyoids had internal cheek pouches.

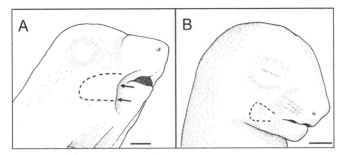

Figure 11. External and internal cheek pouches. A, Developing external cheek pouch in a neonatal *Dipodomys elephantinus* (big-eared kangaroo rat, approximately thirty days of age); arrows point to the anterior opening of the external pouch beside the mouth. B, Internal cheek pouch in a prenatal *Eutamias minimus* (least chipmunk). Dotted lines trace the cheek pouches; scale bars = 2 mm. Redrawn and substantially modified from the photographs in Figure 1 in Brylski and Hall (1988b: 144).

Moreover, the external cheek pouch arose during ontogeny from the buccal epithelium (which lines the internal cheek pouches of ancestral and other rodents). The developmental mechanism is a common one, epithelial evagination, during which the epithelium takes up a new position and participates in new interactions.

In the case of the genesis of the external cheek pouch, the evagination begins at the corner of the mouth, which (uniquely in geomyoids) participates in the evagination; Brylski and Hall showed that the novel external pouch is the result of a small shift in the location and magnitude of the evagination to include the lip epithelium at the corner of the mouth. As the lips develop in tandem with the growth of the snout, the evaginated corner of the mouth is transformed into the opening of the external pouch.

Brylski and Hall speculated, with good reason, that the external pouch was not originally lined with buccal epithelium (as in internal pouches) and then only later became furry; instead, the furriness of the external pouch was the 'direct result of pouch externalization due to an inductive interaction resulting from the novel juxtaposition of the pouch and facial epithelia' (Brylski and Hall 1988a: 394). That such a small change in a developmental mechanism can have such a dramatic effect, coupled with the fact that no living geomyoids have both internal and external pouches, suggests that there is no intermediate ancestor between rodents with internal and rodents with external pouches. (It would be difficult, both developmentally and functionally, to have both internal and external pouches.) So changes in developmental mechanisms may produce coordinated change and thereby participate in the evolution of a lineage; in other words, development may drive evolution by providing the material basis for a new structure. Again, we see the complex interplay between development and evolution and the propriety of an evo–devo explanation.

The Developmental Origins of the Turtle Shell

Two recent publications underscore the crucial role of changes in development in the generation of the turtle shell as an evolutionary novelty; for details of the anatomy of the turtle shell, see Gilbert et al. (2001) and Rieppel (2001). For our purposes, only a few general observations are required: first, the dermal armor of turtles comprises a carapace covering the back of their trunk, and a plastron covering their belly; second, the shoulder blade (scapula) of turtles – uniquely among tetrapods – resides within the rib cage.

New evolutionary developmental studies have mortally weakened the widely held hypothesis that the turtle shell arose gradually, through the accretion of small changes in development, an hypothesis already struggling from its lack of fit with the fossil record and with molecular data. Rieppel (2001) can now claim that the gradualistic model is not compatible with the development of turtles, as shown by Burke (1991) and Gilbert et al. (2001). Gilbert et al. (2001) were able to confirm that the scapula of the turtle develops within the rib cage as a function of a deflection of rib growth to a new position, and this is likely the result of an inductive interaction within the so-called carapacial ridge (CR).

The outer edge of the carapace eventually forms from the CR, which arises in the early embryo dorsal to the limb buds on the lateral surfaces (see Figure 12). Turtles' ribs develop laterally, rather than ventrally, because of the CR; when the CR is either surgically removed or prevented from forming, rib morphogenesis occurs as it does in non-turtle (in fact, non-chelonian) vertebrates (Burke 1991).[8] These are the only two possibilities: either ribs develop deep or superficial to the scapula; moreover, 'there are no intermediates, and there is only one way to get from one condition to the other, which is the redirection of the migration, through the embryonic body, of the precursor cells that will form the ribs' (Rieppel 2001: 991). As Gilbert et al. note, the carapacial ridge forms through the thickening of the ectoderm supported by condensed mesenchyme, which is a typical configuration for epithelial–mesenchymal interaction (2001: 49). The CR is responsible for the redirection of cell migration, as well as the direction of turtle development along a new path.

A simple epithelial–mesenchymal interaction (the same mechanism involved in the previous example) at the onset of carapace development thus may serve as a basis for new hypotheses about the evolution of the turtle body

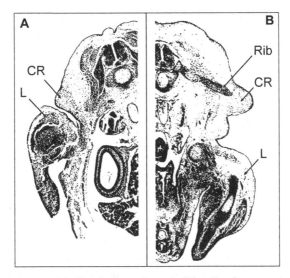

Figure 12. The carapacial ridge in the red-eared slider, *Trachemys scripta*. These are depictions of serial sections of stained embryos through extended limb buds at days 23 (A) and 29 (B) of incubation. (L refers to limb buds and CR to carapacial ridge.) As depicted in A, the carapacial ridge has already formed by day 23, but the rib does not enter into the CR until several days later (as depicted in B). Redrawn and substantially modified from the image in Figure 2 in Gilbert et al. (2001: 50).

plan – particularly, its *rapid* evolution, such that the turtle body plan arose at once rather than gradually and stepwise. Accordingly, the evolutionary biologist's decisions about how to explain evolutionary ancestral relationships and the origins of developmental novelties cannot (always) be made without recourse to the details of development (also see Stern 2000).

AN UNHAPPY SYNTHESIS?

These examples serve to justify evo–devoists' efforts to synthesise development and evolution. It is not always appropriate to presuppose reliable development in a theory of evolution, for developmental mechanisms themselves evolve (or are conserved) through evolutionary time, and evolution and development are mutually constrained by the other. Even if our aim is to understand evolution alone, and development evolves, then evolutionary biology must go well beyond the Modern Synthesis (and it has) in order to explore the phylogenetic implications of development.

In a recent essay, Kim Sterelny (2000b) has suggested that the required changes to Neo-Darwinian evolutionary theory (in particular) are easily made in order to integrate developmental explanations, if not already part and parcel of the Neo-Darwinian perspective. He concludes his article with the following claim: 'I do not see any fundamental conceptual problem for evolutionary biology' on any account of the role of development in evolution that he has surveyed (p. S386). The idea seems to be that, as long as we strive for completeness in our explanations and so recognise that evolution is a two-step process – developmental introduction of variation (e.g., a phenotypic novelty) followed by selection and a change in gene frequency in the population (West-Eberhard 1998: 8419) – then evolution and development are straightforwardly synthesised.

I submit that Sterelny has misspoken here, in referring to the lack of 'fundamental conceptual problems'. Presumably, he means that we need not discard evolutionary theory whole hog just because we need now account for development, its evolution, and its role in explaining particular evolutionary changes. Fair enough. But in accounting for development within evolutionary biology, traditional and Neo-Darwinian evolutionary theory do indeed face some fundamental conceptual challenges.

Consider first Sterelny's argument that, within evo–devo, the focus of evolutionary explanation changes from adaptation as such to evolvability, or the ability of an organism to generate (adaptive) variability (see, e.g., Kirschner and Gerhart 1998). In outlining the nature of the problem of evolvability, Sterelny draws on work in evo–devo on highly conserved elements of development (such as the homeobox genes) and on the phenomenon of modularity. (As already noted, development is modular if traits or trait complexes develop relatively independently of one another.) Sterelny's argument is that if evolvability can be explained, then 'explaining adaptation would be relatively straightforward' (Sterelny 2000b: S377). However, if synthesising evolution and development leads to changing the basic *explanandum* of evolutionary theory – what Sterelny calls 'evolution's "hard problem"' (p. S376) – from adaptation to evolvability, then that is indeed symptomatic of a fundamental conceptual challenge posed to evolution by development.

Consider next Arthur's (2000) manifesto-style suggestion that there are five lacunae in Neo-Darwinian evolutionary theory: (1) it omits all intervening (developmental) steps between mutation and selection; (2) its almost exclusive focus on selection (which is 'destructive') neglects the creative generation of variation (in development and otherwise); (3) it may forswear developmental stability in favour of evolutionary change; (4) in its preoccupation with how organisms respond to environmental problems, it is externalist; and (5) in its extreme formulations, it is exclusively gradualist. Arthur suggests that the turn to evo–devo closes these gaps in two ways: by redirecting attention to what is standardly ignored, assumed, backgrounded, or blackboxed, and, *pace* Sterelny, by revising the core concepts of Neo-Darwinian evolutionary theory. (Insofar as evolutionary theory is not exclusively Neo-Darwinian, some of the required revisions may be minor by comparison, though Sterelny does have Neo-Darwinism in mind.)

Gaps (1) and (2) *may* be filled relatively straightforwardly as Sterelny imagines, provided that evolutionary biologists are willing to open the developmental (epigenetic) box and explore the manifold ways in which ontogeny impinges on phylogeny. This is not a matter of merely admitting that epigenetic processes occur in development but rather of exploring how these processes are crucial to understanding evolution (as in the examples explored herein).

As discussed in previous chapters, Weiss and Fullerton (2000) offer a radical perspective on how a focus on epigenetics may enrich the study of evolution in developmental context. They suggest that we consider 'that it is *not* the genome that is especially conserved by evolution. Suppose the ephemeral phenotype really *is* what we need to understand and what persists over time. Genes would then be "only" the meandering spoor left by the process of evolution by phenotype. Perhaps we have hidden behind the Modern Synthesis, and the idea that all the action is in gene frequencies, for too long' (p. 192; see also Newman and Müller 2000 and Robert 2001a).[9] Because 'evolution works by phenotypes, whole organisms, not genotypes', the Neo-Darwinian account of what evolution *is* would require substantial conceptual overhaul (Weiss and Fullerton 2000: 193).

Filling gap (3) requires attention to the interplay between stasis and change, as well as detailed analysis of developmental constraints biasing phenotypic outcomes (as in the example of the blind cave fish; for more on these themes, see Fusco 2001). Sterelny correctly suggests that developmental constraints already form part of mainstream evolutionary theory, but given Amundson's (1994) discussion of the substantial differences between (developmental) constraints on form (constraints$_F$) and (evolutionary) constraints on adaptation (constraints$_A$), we should be wary of the former being ignored in favour of the latter. Constraints$_F$ are restrictions on possible types of organic form, whereas constraints$_A$ are restrictions on adaptation; constraints$_F$ may result in constraints$_A$, but there is no necessary relationship between the two. In other words, constraints$_A$ are a subclass of constraints$_F$. Evolutionary theory's encounter with constraints$_F$ would surely involve a subtle but substantive transformation in evolutionary theory; but evolutionary theory's encounter with constraints$_A$ generates quite minor changes by comparison. I have seen little evidence that many evolutionary theorists have fully digested constraints$_F$.

Sterelny himself speaks of constraints on variability, a formulation which threatens to collapse Amundson's useful distinction (see also von Dassow and Munro 1999: 312). (It is worth noting that the long-term stasis implied by Eldredge and Gould's theory of 'punctuated equilibrium' [1972] is a nice example of how modern evolutionary theory already accounts for evolutionary [and developmental] stability – but the theory still tends to generate hostility, though Dawkins [1986] and Dennett [1995] both believe it is part and parcel of the Neo-Darwinian synthesis.)

The fourth and fifth gaps may be more problematic for Neo-Darwinians. Arthur's version of evo–devo corrects for gap (4) by redirecting evolutionary attention to both the insides and the outsides of organisms. The usual Neo-Darwinian story is that 'the environment imposes a set of adaptive demands on a population, and selection shapes that population so that it meets those demands increasingly well' (Sterelny 2000: S372–S373). This Neo-Darwinian picture of mutation–variation–selection is externalist (even though mutations are surely internal!): successful variants are those that respond well to external pressures exerted by environments.[10] However, this picture ignores the developmental intermediaries between mutation, the production of variation, and the sieving process of selection, and it also presumes a priori that genetic mutation is the ground of all evolutionary change. In contrast, Arthur (1997) and Fusco (2001), for instance, maintain that an organism's internal structures and developmental interactions may be positively favoured by selection. Evolution is not exclusively about how well organisms fit external environments which putatively pose problems for organisms to solve, but also how well an organism's insides fit together: how well it is internally integrated (which may or may not assist in responding well to external pressures). Darwin knew this, as did Fisher, Wright, and Haldane, but present-day Neo-Darwinians know it not – the fallout of blackboxing development, in fact, of blackboxing the organism (Shishkin 1992). Even should the black box be opened, overcoming the externalism of evolutionary theory will represent a challenge; at the least, it will require reinterpreting fundamental evolutionary concepts. Shishkin (1992: 37) goes further still, urging that we understand evolution as 'a transformation of integral properties of the developmental system' in establishing developmental stability.

What of the fifth gap? To Neo-Darwinians, this putative gap is in fact a virtue. The Modern Synthesisers were, of course, expressly committed to gradualism: evolution occurs by the accretion of minute adaptive changes, and speciation occurs by mega-accretion. However, many evolutionary biologists (those who would consider the Neo-Darwinians to be on the extremist fringe) already acknowledge that there are exceptions to gradualism, as in the evolution of basic body plans (and not only in the case of turtles – see, e.g., Raff 1996; Arthur 2000: 55; Arthur 1997; Robert 2001a). A shared objective of evolutionary developmental biologists and many evolutionary biologists proper is to secure a firm place in evolutionary theory for such putative exceptions. For evo–devoists, the question of the nature of mechanisms of macroevolution is an open one; Stern (2000) has recently shown that analysis of development may be ineliminably important in deciding between competing evolutionary hypotheses regarding micromutational versus macromutational evolution. Focusing on evolvability should be of distinct usefulness here.

Given the conceptual retooling required to fill these lacunae, and the changes in theory building and experimental design necessary to fully realise the promise of evo–devo, development surely does pose challenges to (at least Neo-Darwinian) evolutionary theory. Sterelny's version of evo–devo attends insufficiently to the fundamental problems of development (differentiation, growth, and change) in relation to evolution. If, however, we engage a broader account of evo–devo, then we can avoid Sterelny's conclusion.

HAPPINESS EVER AFTER?

In the effort to reorganise relationships between disparate disciplines, at least four strategies are possible: assimilation (or subsumption), fusion, contamination, and synthesis.[11] Contamination (which is not to be understood pejoratively) is unavoidable, as witnessed by the past 100 years of the history of biology, wherein genetic and evolutionary perspectives have permeated every biological subdiscipline. The choice, then, is amongst fusion, assimilation, and synthesis. Some theorists might favour fusion, according to which both disciplines lose their distinct identities and meld together seamlessly. However, such fusion may be practically impossible or, if possible, then remarkably impure, tending too much toward assimilation – the aim of adherents to the modern consensus. As against both of these options, I prefer the fourth strategy in order to guard against the pauperisation of development under the aegis of genetics and evolutionary theory.[12]

Synthesis stems from the inevitability of contamination but ensures its reciprocity. In so doing, it avoids the assimilation characteristic of biology throughout the twentieth century, yet it also avoids fusion's melting-pot mentality, permitting the emergence of new approaches and techniques as well as the division of biological labour into differentiative and integrative problem sets. Both of these are crucial in understanding and explaining developing organisms, but only when they condition each other.

My contention is that evo–devo is as successful as it is fundamentally because it is a *synthesis* – it does not attempt to accommodate or subsume development within evolutionary biology or vice versa; rather, it brings developmental and evolutionary biology together in a new discipline. Thus evo–devo does not spell an end to either evolutionary or developmental biology proper but makes room for them to interact fruitfully, even synergistically, and to uncover phenomena inaccessible to either evolutionary or developmental analysis alone.

But considerable work remains to be done. Particular research programmes within evo–devo may be overly friendly to the modern consensus and therefore subject to challenge on the basis of the considerations of the preceding chapters. Accordingly, despite my intimation of 'happiness ever after', the view of evolutionary developmental biology promulgated here is not unassailable, as I show in the next chapter.

7

The Devil is in the Gestalt

Theorists are exasperated to be told what they have 'always known'.
Yet there is a difference between knowing in a parenthetical, 'of course
it's important' way about the intimacy and reciprocality of organism-
environmental exchanges in development and evolution, say, and in-
corporating the knowledge in models and explanations, research and
theory.

– Susan Oyama (2000a)

How we understand both heredity and evolution depends crucially on how we
understand development. Accordingly, theories of evolution and of development are critically interdependent. Those endorsing an evolutionary theory ignorant of development – or an account of development ignorant of evolution –
have enjoyed centre stage for most of the past century. An insurrection is long
overdue.

Although evolutionary developmental biology, in its various formulations,
represents a most promising synthesis of development and evolution, there
are alternative proposals currently in circulation. In this final chapter, I explore one such alternative in detail – the developmental systems perspective. I
highlight its benefits and limitations as compared with evolutionary developmental biology as a theoretical, empirical, and methodological framework for
a genuinely synthetic biology comprising genetics, developmental biology,
and evolution. But first, I explore how the modern consensus might mislead
us in the project of synthesising biology.

STANDARD VIEWS

A rough taxonomy of some standard positions on the relationship amongst
genetics, developmental biology, and evolutionary biology will set the stage

for the discussion in this chapter. They tend to fall into three classes: (1) gene-centric approaches, according to which considerations about gene frequency and gene expression are all that are required to successfully answer evolutionary and developmental questions; (2) the genes-in-context approaches characteristic of evo–devo, which extend the scope of (1) by focusing additionally on epigenetics and the evolution of developmental processes; and (3) non-gene-centric approaches, such as the developmental systems perspective, which refocus both evolutionary and developmental inquiry on both genetic and generic factors, forces, and mechanisms.[1]

The first approach is characteristic of those enamoured of the modern consensus, whereby the gene is the basis of both development and evolution. An exemplar of this first approach to the relationship amongst genetics, development, and evolution is Jeffrey Schwartz. Recall the example of the homeobox genes discussed in Chapter 2. Schwartz argues that homeobox genes are pivotal in individual ontogeny, productive of both normal development and also, when their timing is off, of monstrous macromutations; Schwartz speculates, in the absence of solid evidence, that these mutations might silently accumulate and then eventually be expressed in sufficient numbers such that several monsters produced all at once could interbreed and this, in time, could lead to speciation.[2] Schwartz contends that in ontogeny, 'all that is necessary is that homeobox genes are either turned on or they are not' at the appropriate time (Schwartz 1999: 362, 368–369).

Schwartz could have suggested, more plausibly, that homeobox genes are one of many factors in the production of large-scale morphological changes at the level of organisms, and that the organismal level is the level at which selection pressures are operative in the establishment of new species. Such a view would have been congruent with the perspective of evolutionary developmental biologists that variation between individual organisms is introduced ontogenetically, as a result of genetic–epigenetic–phenotypic–environmental interactive processes. Instead, Schwartz opts for an implausible suggestion: homeobox genes 'control everything' and 'run the whole show'. Therefore, 'the morphologies that make up an organism ultimately derive from the turning on and off of homeobox genes'. According to Schwartz, then, 'timing is everything': the timing of homeobox gene expression makes all the difference between eels and elephants, flies and frogs, mice and men (Schwartz 1999: 36, 34, 44, 280).

Graham Budd complicates this sort of story by noting some of the other transformations that would be required for a change in the timing of homeobox gene expression to have an evolutionary, or even an ontogenetically functional, impact. One of the examples he discusses is the feeding appendages of crustaceans; these are, like the vast majority of functional features, intricate and well-integrated parts of the organism. If it were shown that an alteration in the timing of homeobox gene expression results in a homeotically transformed feeding appendage, an outstanding problem would remain, namely that of 'the integration of the new morphology into the functional complex that is an animal as a whole'. Not only would the new appendage need to be integrated with the other feeding appendages, so too would necessary correlated alterations in muscles and in the nervous system.[3] In other words, Schwartz's story ignores integration, a key element of any serious account of organismal development. (Budd's model of homeotic takeover, discussed in Chapter 2, is particularly sensitive to the concern for morphological and functional integration.)

Schwartz's problem is that he adopts a modern consensus view of development, according to which genes are foundational and the only foci of developmental interest; epigenetics, for Schwartz, is no more than the differential regulation and expression of homeobox genes; therefore, his is a genes-*plus* account of ontogenesis, according to which development is subsumed under genetics. Genes and phenotypic traits are tightly linked on this view, and evolution for Schwartz is exclusively a matter of gene frequencies in populations; hence his peculiar account of evolution.[4]

Genes-in-context approaches, or approaches of the second type, may dispense with the thesis of genetic animism but otherwise hold, to varying degrees, to the other elements of the modern consensus. A good exemplar here is the version of evolutionary developmental biology advanced by Brian Hall. Hall agrees with mainstream evolutionary theorists that the gene is 'the unit of transmission of heredity' (Hall 2000a: 177), and also that 'the genetic basis for development lies preformed in the DNA of the egg and subsequently in the zygote'. However, there are non-genetic preformed structures as well, including the egg's cytoplasm and organelles, and although Hall allows that epigenetic events direct developmental processes, he holds that 'it is a mistake to speak of epigenetics as nongenetic or of genetic versus epigenetic factors as if one is always in the ascendancy or acting to the exclusion of the other'. For Hall, 'epigenetic control' simply is 'control of gene expression' (Hall 1999: 113, 114).

Nevertheless, Hall's account of evo–devo is by no means exclusively or even primarily gene-centric, as he sees the *cellular context* of gene action and activation as centrally important in evo–devo. Whatever developmental information is contained in the genes requires a cellular and in fact an extracellular matrix for expression; moreover, some developmental information simply is not to be found in the genome – at least, not in the genome of the developing organism. Some of it is to be found in the maternal genome and so is inherited genetically (Hall 1998: 203);[5] some of it is to be found in the nucleolus and ribsomes (Hall 1999: 113). The core of evo–devo, then, is the epigenetic construction of the developing individual from genetic and other raw materials in the egg. Furthermore, although the gene is the unit of hereditary transmission, it is 'cells and their immediately adjacent peri- and extracellular matrices' that 'carry out the selective responses that allows [sic] organisms to develop, adapt to their environment, modify their development, and translate the effects of gene mutations and genetic assimilation into evolutionary change' (Hall 1999: 400). Hall is therefore a pluralist regarding the fundamental units of evo–devo; individual cells and cell condensations (Hall and Miyake 1992, 1997, 2000; Hall 2003) are at the core of evo–devo, even though genes are the units of transmission of heredity and primary suppliers of the raw materials of ontogenesis (see also Robert et al. 2001).

Of course, it might be objected that Hall's approach is nonetheless excessively genecentric. An alternative perspective within this second group is that offered by Scott Gilbert, John Opitz, and Rudy Raff (1996). They derive their inspiration from a now-almost-defunct experimental embryology and so seek to resurrect the notion of a *morphogenetic field* as a cornerstone of evo–devo. Gilbert et al. (1996) remark that the morphogenetic field was 'one of those notions that was so powerful as to be assumed rather than continually proven', and it served as the basic explanatory concept – and entity – in pre-molecular embryology. The morphogenetic field is a modular, physical web of embryological inputs defining cells and delimiting their interactions. There are eye fields and limb fields and heart fields, for instance, comprising and regulating particular collections of cells required for the morphogenesis of eyes, limbs, and hearts. Once upon a time, these fields were 'innocent of genes', but now, in the genetic era, that is no longer the case. Gilbert et al.'s interpretation permits fields to be genetically defined, but because the field is (and always was) intended as competition to the gene's eye view of development, their perspective is not gene-centric. In other words, although genes remain important for Gilbert et al., genes simply do not control ontogeny (morphogenetic fields take over that role in a newly synthesised biology); nor are genes methodologically central a priori (Gilbert et al. 1996: 359, 367; see also Gilbert and Sarkar 2000).

However, there is no unequivocal reason that morphogenetic fields be genetically defined. The genes-in-context trope may be taken much further – so far, in fact, as to constitute a third set of synthetic efforts. A primary instance here is the case of the developmental systems perspective,[6] a theoretical perspective advanced by Susan Oyama, Gilbert Gottlieb, Paul Griffiths, Russell Gray, and others (see Oyama et al. 2001), though they had numerous precursors.[7] Here I shall focus largely on the work of Oyama.

NONSTANDARD VIEWS

Oyama's synthesis of development, genetics, and evolution, introduced in her 1985 book *The Ontogeny of Information*, displaces the idea that genes are either programmes or blueprints for development – in effect, she rejects all three theses of the modern consensus. According to Oyama, genes must be *deeply* contextualised. 'If development is to reenter evolutionary theory, it should be development that integrates genes into organisms, and organisms into the many levels of the environment that enter into their ontogenetic construction' (Oyama 2000a: 113). Developmental systems theorists reject dichotomous views of development which partition ontogenetic causes into genetic causes and generic ones (everything else, but usually mainly environmental causes). For Oyama, as for other adherents to DST, developmental information does not pre-exist individual ontogenies but rather emerges from the *interactions* of dispersed developmental resources of various kinds – hence, the *ontogeny* of information. In contrast with the mainstream interpretation of heredity as transmission of genetic information between generations, developmental systems theorists underscore the construction of developmental information in each generation from a range of resources. Consequently, developmental processes both generate the relatively reliable reproduction of type and also introduce variation of potential, eventual evolutionary significance.

DST has been deployed to dissolve the traditional nature–nurture dichotomy in biology and psychology and to underscore an alternative to the gene's-eye view of evolution and development.[8] The resultant proposed synthesis of evolutionary biology, developmental biology, and genetics occurs without recourse to the 'hegemony of the gene' (a phrase used by Falk 1991: 470), and therefore it represents something other, something much more revolutionary, than a refinement or extension of the modern and interactionist consensuses.

Recall the claim in Chapter 6 that evo–devo is preferable to the gene's-eye view of development and evolution; I now elaborate developmental systems theory in order to ascertain its conceptual and methodological relationship to evo–devo. Although Kim Sterelny has suggested that the prospect for a successful synthesis of evolution and development 'does not stand or fall' on any of DST's 'distinctive theses' (Sterelny 2000b: S384),[9] adequately assessing this judgment requires sustained exploration of these distinctive theses.

Note that much of what developmental systems theorists have to say is not overly original, for some of their core ideas are commonplace amongst biologists (at least in theory if not in practice); but the particular *coalescence* of ideas and strategies within DST, the remarkable way in which DST brings together experimental and theoretical traditions from biology and comparative psychology, is both impressive and worth exploring in detail as a possible companion or perhaps even, though less plausibly, as an alternative to evo–devo. The following presentation of DST is admittedly charitable, resulting from both the benefit of hindsight (that is, of having digested the potent critiques of Sterelny and others[10]) and also the desire to salvage the most compelling elements of this perspective from those who might dismiss it as pseudo-philosophical hogwash. Having discussed DST with a number of biologists in recent years, almost all of whom appreciate various aspects of DST but none of whom would refer to themselves as DSTers, I find that such a charitable reading as mine is essential in order that DST have an opportunity to influence the progress of biology.[11]

For Oyama, at least, the central, and fundamental, construct of DST is the *developmental system*.[12] A developmental system is 'a mobile set of interacting influences and entities' comprising 'all influences on development, at all levels of analysis', including the molecular, cellular, organismal, ecological, and biogeographical: 'the developmental system includes not only the organism but also features of the extraorganismic environment that influence development' (Oyama 2000a: 72, 82). These organism–environment systems host a complex of 'more or less reliably occurring cascades of developmental contingencies'; the intrasystemic interactions 'singly may not be universal or necessary, but . . . can nevertheless produce very reliable consequences because of their interrelations' (Oyama 1999: 189). The interactive resource matrix comprising the developmental system is contingent and may be discontinuous in both time and space, but the components of the matrix share the joint developmental and evolutionary task of reliably (though not unfailingly) reproducing the organism–environment dyad.

That DST is not a specific theory, and that not all DSTers adopt an identical stance, forces the imaginative reconstruction of the distinctive theses of developmental systems approaches. Therefore, it is noteworthy that two recent, independent accounts of the interrelated theses central to DST (Robert et al. 2001 and Oyama et al. 2001) have converged on an almost identical set of them. (Here I borrow the terminology of Robert et al. with Oyama et al.'s usage in parentheses.)

Contextualism (context sensitivity and contingency): Whether the resources for development come together as required (in the right way at the right time) is a contingent affair. 'The reliably present, overdetermined, multilevelled context of development is sufficient to explain the remarkable reliability of reproduction and development, without invoking the problematic notion of a genetic program' (Robert et al. 2001: 955). Moreover, as the discussion of creative development in Chapter 5 suggests, the reliable presence of developmental resources is in large part the result of their having been constrained, influenced, selected, and constructed by organisms, and their conspecifics and symbionts. Oyama et al. (2001: 3) note that the persistent use of the metaphor of specifically genetic information helps to perpetuate the myth that context sensitivity and developmental contingency are mere obstacles (or noise, or interference) to be methodologically overcome or filtered out in the effort to understand how genes cause phenotypic outcomes. Considering the problems with the metaphors of genetic information, instructions, and programmes explored in earlier chapters, developmental systems theorists, in conjunction with evo–devoists, emphasise the need to investigate actual developmental processes. According to DST, then, contingency and context are decisive, not digressions from the causal truth.

Nonpreformationism (development as construction): Developmental systems theorists reject the common claim that preformationism and epigenesis may be reconciled according to the modern consensus, namely that preformed genetic information is expressed epigenetically. DST offers a 'thoroughly epigenetic account of development' according to which developmental information – in genes, genomes, cells, environments, or elsewhere – emerges during development rather than being preformed and transmitted between generations (Oyama et al. 2001: 4). DST is therefore sometimes referred to as 'developmental constructionism'; Oyama has switched from using 'interactionism' to using 'constructivist interactionism', reflecting the perspective that developmental interactions generate new information in ontogeny rather than merely triggering the putatively specific information contained only in genes (Oyama 2000a, 2000b). Robert et al. (2001: 955) note that there are both strong and weak versions of nonpreformationism amongst DSTers: according to the strong version, no developmental information whatever is preformed (and there may not even be a scientifically respectable account of biological information to be had); according to the weak version, developmental information may be preformed, but it exists in a wide range of resources (not just genes). In either case, DST denies a unique informational role for genes.

Causal co-interactionism (joint determination by multiple causes): Constructive causal interactions in development are not exhausted by gene

activation but rather involve 'inducing, facilitating, maintaining, and participating in time-sensitive positive and negative feedback loops at a variety of
levels within and without the developing organism' (Robert et al. 2001: 955).
The constructive causal interactions comprising organismal development are
complex; moreover, their effects are not simply additive. Consequently, a
systems account of causality cannot be reduced to the formula of 'genes-
plus-(stimuli, trigger, or other cause)' (see Chapter 4), and in many cases it
is implausible either 'to assign causal primacy [or] to dichotomise developmental causation into internal and external components' (Gray 1992: 175).
This is not to say that all sources of causality play identical or equivalent or
equally important roles, but only that whatever differences exist between their
roles 'do not justify building theories of development and evolution around a
distinction between what genes do and what every other causal factor does'
(Oyama et al. 2001: 3).

Causal dispersion (distributed control): The first and second theses suggest that causal power is not centralised; the third thesis suggests that it not
be dichotomised, either; consequently, for DST, causal power is dispersed
throughout the developmental system. Evo–devoists such as Gabriel Dover
have argued that some aspects of development, such as cell–cell signalling,
cannot be represented as simple causal pathways but rather should be construed in terms of networks of causal interactions (see, e.g., Dover 2000: 1156;
see also Solé et al. 2000). Causal power is not contained within any particular
entity or class of entities but rather resides in the contingent relations between
developmental interactants within such networks. According to DST, then,
'a gene is a resource among others rather than a directing intelligence that
uses resources for its own ends' (Oyama 2000a: 118). This thesis of DST thus
refocuses attention away from the perspective that genes are ontogenetically
(and ontologically) primary and toward a multiplicity of factors, forces, and
mechanisms operative in, and constitutive of, development.

Expanded pool of interactants[13]: According to the interactionist consensus, genes and environments interact in the production of traits. As already noted, DST rejects this particular blanket dichotomy of developmental
resources–causes–factors into genetic and generic varieties. There are more
than just two types of interactants amongst the heterogeneous components
of a developmental system, and a multiplicity of ways in which they interact
in development. Within the organism, interactants include DNA sequences,
mRNA, cells, the extracellular matrix, hormones, enzymes, metabolites, and
tissues; beyond the organism, some exemplary developmental interactants
are aspects of the organism's habitat (including temperature and nutritional
resources), the organism's behaviour and that of conspecifics and others

(Gottlieb 1992, 1997; Johnston and Gottlieb 1990), social structure (Keller and Ross 1993), and (depending on the system) even gravity and sunlight (van der Weele 1999; Gilbert 2001). Recognition of these interactants is not, of course, the exclusive domain of DST, but DST demands that we explore the specific nature of constructive interactions between all developmental resources as part of any adequate account of organismal development.

Extended inheritance[14]: According to both the modern consensus and the account of evolutionary developmental biology surveyed here, the sole unit of hereditary transmission is the gene (though biologists of course recognise that complex cellular structures are also inherited). As Gray notes, 'such is the dominance of genocentric thinking in biology that the claim that only genes are inherited seems like a simple truism – a statement of accepted fact rather than a contestable theoretical position' (Gray 2001: 194). Thus do developmental systems theorists challenge this seeming truism, urging instead that there is considerably more to inheritance than just these items, including all reliably present elements of the developmental context.

As Dover (2000: 1154) notes, 'DNA is a far more unstable molecule, on an evolutionary scale, than is conventionally thought'. However, whatever stability genetic inheritance enjoys, even on a much shorter timescale, depends critically on the inheritance of those resources that are part of the expanded pool of interactants. Oyama et al. (2001: 4) note that 'some of these resources are familiar – chromosomes, nutrients, ambient temperatures, childcare'; less familiar inherited resources include the chromatin marking system, chemical gradients in the cytoplasm, and the altered environments (and associated altered selection pressures) generated through niche construction. Accordingly, DST proposes a broad interpretation of inheritance and, more basically, reinterprets hereditary transmission as contingent but reliable reconstruction of resources-in-interactive-networks in the next life cycle (see Oyama 2000a: 199).

Evolutionary developmental systems[15]: Given this broad account of inheritance as construction, evolution works on elements at all levels of developmental systems. In other words, 'selection pressures act on the whole developmental manifold at all levels of complexity' (Robert et al. 2001: 956).[16] According to DST, then, evolution should be defined as change in developmental systems – change in the life cycles of organisms in their co-constructed niches – tracked by differential reproduction and distribution of developmental systems.

In summary, according to a developmental systems view, genes are but one of many inherited developmental resources; these resources cannot be dichotomised into (specific) genetic and (non-specific) generic classes; DNA sequences and other resources participate in complex, non-additive, time-sensitive, constructive networks of interactions, such that control of development is dispersed; accordingly, causation must be tracked in multiple directions; evolution occurs through changes in organism–environment systems, reflected in their frequency and distribution; and so understanding both development and evolution 'requires "unpacking" the developmental system' (Oyama 2000b: 180).

DST, EVO–DEVO, AND THE MODERN CONSENSUS

It is evident that DST rejects all three theses of the modern consensus, not only as stated but also in principle. Let me rehearse the reasons why, referring where appropriate to evo–devo to reflect some contrasts between these projects. First, recall that genetic informationism is the thesis that genes contain the entirety of the preformed, species-specific evolutionary and developmental information. DST rejects two components of this thesis, namely that DNA is all encompassing and that genes are preformed.

That DNA does not contain *all* of the relevant developmental and evolutionary information is, I think, widely agreed upon. Evo–devo, for instance, could conceivably leave room for epigenetic inheritance systems relatively independent from genetic inheritance systems, or perhaps for the recognition that a kind of genetic or epigenetic systemism is in order (as in Wolffe 1998).

DST takes this a step further, by underscoring that much more than DNA is inherited at conception, birth, and beyond. In this regard, DST rejects not only the thesis of genetic informationism but also that of genetic primacy, according to which the gene is the unit of transmission in heredity. DST, as I have noted, understands 'inheritance' more broadly than does evo–devo, and therefore it parts company with evo–devo on the issue of the putative evolutionary primacy of the genes (see, e.g., Jablonka and Lamb 2002).

The second aspect of genetic informationism rejected in principle by DST is that genes are informationally preformed. Evo–devoists are more circumspect; whatever developmental information is contained in the gene requires a cellular and in fact an extracellular matrix for expression; but, nevertheless, genes contain basic developmental information. Developmental systems theorists break with this point of view in two related steps. First, there is a problem of containment. According to DST, DNA does not *contain* preformed developmental information, awaiting epigenetic release; as Russell Gray puts it, 'developmental information is not *in* the genes, nor is it *in* the environment, but rather it develops in the fluid, contingent *relation* between the two' (Gray

1992: 177). Developmental information, including information to be gleaned from DNA, itself has an ontogeny.

The second step, taken by some but not all developmental systems theorists, is that not only is the function of DNA dependent on cellular and extracellular context, but so too is the very structure of DNA. Considering this second step takes us from genetic informationism and genetic primacy to the related thesis of genetic animism. The idea here is that the ontogenetic structure and functional significance of genes – and, in fact, genes themselves – are co-constructed with and by a full range of molecular and non-molecular developmental resources. Such a viewpoint is virtually indistinguishable from my account of constitutive epigenetics in Chapter 4. Not only does genetic information but so too do genes themselves have an ontogeny, in both a functional and a structural sense. On this view, genes do not precede development but are rather constructed during development. Consequently, genes, defined as particular, particulate stretches of DNA, can be neither the units of hereditary transmission, nor, as the modern consensus would have it, developmental prime movers.[17]

If my account of constitutive epigenetics is even partly right, then there is a drastic difference between evo–devo and DST; even though both would drop the thesis of genetic animism, DST goes much further than evo–devo in deeply contextualising both the function and the structure of DNA. But if genes must be reconceived in such a Draconian manner as contingent, constructed templates rather than as stable ontogenetic or evolutionary entities, perhaps we would do well to stop talking about genes in development altogether and instead focus on the generative processes of development as such.

CHALLENGING DEVELOPMENTAL SYSTEMS

That is, to be sure, a radical conclusion. How plausible is it? There are a number of ways in which to address this question. First, I will describe and critically assess a recent challenge to DST issued by Kenneth Schaffner. Then I will turn to the larger challenge of investigating whether DST could make a practical difference within biology by, for instance, generating research programmes, supplying models, or even just providing some tools for model building. Finally, I will attend to the prospect of DST's conceptual and theoretical impact on the interpretation of experimental results, even where DST provides no practical guidance in the design of research programmes.

Schaffner has explored how core aspects of developmental systems theory play out in a wet-bench setting; he offers a detailed analysis of some laboratory work on the behaviour genetics of the nematode worm *Caenorhabditis elegans*. Schaffner's discussion is enlightening, not least because he unexpectedly uncovers so much complexity in the development of *C. elegans*, a very simple experimental system.[18] However, *C. elegans* is not a good test case for developmental systems theory: 'the very richness of life that the Developmentalist Challenge claims engenders diversity have been hunted down and eliminated from *C. elegans*', a model organism specifically constructed 'to show that the basis for behavior lies in the genes' (Gilbert and Jorgensen 1998: 259–260). So, although perhaps the developmentalist challenge may be overstated in some ways in particular laboratory contexts where genes are easily manipulated in highly derived model systems against a constant background of non-specific enabling conditions, developmental systems theory may be on target in less contrived circumstances.

Having distinguished five putative theoretical commitments of DST, Schaffner (1998) concludes that, although two of these theoretical challenges (nonpreformationism and contextualism) are tolerable, they are also widely accepted independently of DST; meanwhile, the other three challenges issued by DST are either overstated (causal parity) or misguided (indivisibility and unpredictability).

In their response to Schaffner, Griffiths and Knight (1998) maintain that the ostensibly widely accepted theses are not taken seriously enough, even though granted by all concerned. Moreover, they urge that Schaffner's characterisation of the parity thesis is a straw position, one not actually subscribed to by developmental systems theorists. Schaffner (1998: 234) attributes to DST the perspective that genes and other developmental resources are on a par causally, epistemically, and heuristically. He grants that the parity thesis may be true causally (though he holds as well that DNA has special informational priority on the basis of the central dogma of molecular biology). However, Schaffner contends, as noted in Chapter 1, that the parity thesis is clearly false regarding epistemology and methodology. Schaffner at times appears to imply that DSTers hold that all developmental resources are of *equal ontogenetic importance* – whatever that might mean. (See the more plausible interpretation already discussed as part of the thesis of causal co-interactionism.)

The final two putative theoretical commitments of DST that Schaffner discusses – indivisibility and unpredictability – are more problematic. The idea of indivisibility is, according to Schaffner, the position that the effects of genes and environments cannot be analytically separated because they are 'a seamless unification, an amalgam' (Schaffner 1998: 233). Judging from my discussion herein, this is not a central thesis of DST, though the context dependency of development (granted by biologists generally) does imply that the a priori assumption of additivity is suspect, and also that genetic and generic causes cannot be *meaningfully* separated *independently of the context of their interaction.*

A strong commitment to unpredictability, finally, is not necessarily part of DST's challenge to biological practice. Schaffner contends that the unpredictability of DST implies that 'from total information about genes and environment, we cannot predict an organism's traits' (Schaffner 1998: 233). Schaffner denies any strong version of this claim – as do Griffiths and Knight (1998: 257), who are not concerned with whether traits can be predicted, but only *what* they can be predicted *from*. I am less sanguine.

Some non-DST biologists, in an antireductionistic vein, insist on a degree of both unpredictability and inexplicability (Gilbert and Sarkar 2000; Schlichting and Pigliucci 1998). It is worth quoting at length the remarks of Schlichting and Pigliucci:

> The limits that we are trying to outline here are part of the never-ending debate between reductionist and holist philosophies in the quest for an understanding of the natural world. It can be argued that the impressive progress of the most reductionist of the biological sciences, molecular biology, is in fact helping to reinforce a scenario proposed long ago by the holist camp – the essence of a biological system is in the emergent properties of its interacting component parts. We can dismantle the system piece by piece, but the more we do that, the more we realize that these emergent properties can only be investigated when the parts are together. In more pragmatic terms, this is an old problem in mutagenesis studies. For example, we will never be able to uncover all the important genes contributing to the normal development of an embryo, simply because the mutations of many of these genes are lethal, precluding the study of their phenotypic effects. In other cases, redundancy of function also masks the true nature of the mutation. In a metaphorical sense, this is similar to attempting to understand how an automobile works by taking it apart, impairing one major function at a time. For example, we might surmise that the loss of directional ability caused by the 'steering wheel-less' mutant is a fundamental mutation early in the guidance system pathway of the car. However, from our knowledge about automobiles, we know that the steering wheel is actually the terminal component, and we have not really untangled any of the actual complexity. And a car is orders of magnitude less complicated than even the simplest living organism. (Schlichting and Piglincci 1998: 253–255)

The methodological lesson to be gleaned from these comments on reductionism and holism is one we have already encountered in this book: namely, that methodological systemism is required in order to adequately and effectively investigate complex biological phenomena (Riedl 1977; Shishkin 1992; Hall 1999; Dover 2000; Newman and Müller 2000; Solé et al. 2000). Moreover, here we yet again witness the need to focus closely, beyond the metaphors, on the details and dynamics of interactive, emergent, developmental processes.

These two lessons are certainly not unique to developmental systems theory. In fact, they form part and parcel of at least some versions of evolutionary developmental biology, and of what Adam Wilkins has enchantingly called 'enlightened developmental biology' (personal communication, May 2001). So what is the 'cash value' of developmental systems theory? How could it make a difference in practice to working biologists (Kitcher 2001)?

Gray (2001: 202–203) has provided a sketch of DST-inspired research programmes, including the following:

1. Treat pseudo-explanatory claims about genetic programmes or 'genes-for' as potential research questions. It is not enough to conclude that genes function in context – DST demands answers to these questions: which context? how? what is the precise nature of the interactions from gene to phene? are these interactions spatiotemporally dependent? which interactants are involved, and where do they come from? what are their downstream effects?

2. Study extragenetic inheritance both in itself (its existence, longevity, and fidelity) and in a comparative sense (its adaptive value); also explore the co-evolution of genetic and extragenetic inheritance.

3. Study niche construction through field experiments and model building.

4. Explore the relationship between developmental integration and developmental modularity, as well as interactions both within and between modules.

However, as Robert et al. (2001: 959) have maintained, these research programmes are already underway in evo–devo and ecological developmental biology, and so they do not depend on the distinctive theses of DST for their design or execution. It appears, then, that Sterelny's conclusion that DST does not contribute specifically to the synthesis of developmental and evolutionary biology may be correct. But let us not concede too soon.

Consider just the fourth of these research programmes, that dealing with integration and modularity in development. At present, the literature on modularity is – not to put too fine a point on it – messy. Aside from the very few rigorous models of developmental modularity in circulation, such as Atchley and Hall's (1991) model of 'fundamental developmental units' in the morphogenesis of the mammalian dentary, 'developmental modularity' is more of a buzzword and umbrella concept – perhaps more of a metaphor – than it is a technical notion describing a mechanism (or set of mechanisms). That is, although operational definitions of modularity are invoked in specific experimental contexts, no adequate formal or theoretical definition has any widespread currency. Jessica Bolker (2000) has suggested that this is to be expected, inasmuch as we have no general, unifying theory of development from which to derive a theoretical definition. Accordingly, the most we can hope to generate are definitional *desiderata*.

Bolker offers five: (1) the definition should be prospectively operational-isable or forward-looking – that is, we should be able to identify modules on the basis of the definition rather than merely apply the definition only once the development of a particular structure is well understood; (2) it should be applicable across levels of biological organisation and the definition should capture three aspects of modules already recognised in the context of working definitions: (3) modules are internally integrated; (4) their internal integration suggests that modules are emergent individuals; and (5) modules participate with other entities from which they are distinct (Bolker 2000: 773).

Although formal definitions have been proposed, none is widely agreed upon, and none as yet fulfils these five *desiderata*. Dover, for instance, has provided a broad definition of a module as 'an independent unit or process or function that may interact in a variety of combinatorial interactions with a variety of other units or processes or functions' (Dover 2000: 1155). The requirement of independence suggests that the internal rather than the external organisation of the module is a crucial feature, whereas the emphasis on interactions suggests that its external relations are also important to recognise. Ontologically, though, whereas presumably such modules could exist at a variety of levels, if units or processes or functions could all count as modules, then 'module' appears to be a catch-all concept. Moreover, Dover provides no solid justification for considering each of these types of entity as modules and offers no grounds for individuating them.

In contrast, von Dassow and Munro have suggested that 'a developmental module is a collection of elements whose intrinsic behaviors and functional interactions yield a mechanistic explanation of an identifiable developmental process or transformation' (von Dassow and Munro 1999: 313). Again, this conception of modules emphasises both interactivity and internal integration, but here it appears that only structures (not processes) of some sort or an-other would count as modules. Moreover, whether they count as modules is exclusively determined by pragmatic epistemological and methodological considerations. Von Dassow and Munro indicate their awareness that this conception of a module applies only to explaining a process rather than to the nature of the process itself, but they are not bothered by this limitation (p. 313, note 5).

Although they do not actually provide a precise definition of a module, Gilbert and Bolker (2001) do indeed include processes – such as cell–cell signalling – as modules and provide a strong argument as to why this should be the case. They argue that, ultimately, development is a spatiotemporally extended process, and that 'the most significant features of embryos are not structures. Rather, they are the processes and changes embryos undergo, and the mechanisms by which those changes occur' (pp. 9, 10). A consequence of this focus on processes is that we are invited to envision genes as interactants in a process, or elements of a developmental pathway, rather than as independent agents. Gilbert and Bolker emphasise that modules may change without affecting other modules and also may be co-opted to new functional roles. The emphasis here on individuation and interactivity (and so changeability) may come at the expense, however, of the internal integration of a module (pp. 2, 10).

Meanwhile, Bolker's own account of modularity (Bolker 2000) appears at times to de-emphasise interactions between modules in favour of the independence–individuation and internal integration criteria. For Bolker, interactivity may seem an afterthought, and so her account of modularity may be overly atomistic.

Atchley and Hall may be seen to err on the other side. They propose a definition of fundamental developmental units (substitute: modules) as 'those basic structural entities or regulatory phenomena necessary to assemble a complex morphological structure' (Atchley and Hall 1991: 772). Hence, interactivity is essential, whereas there is considerably less emphasis on the internal integration of modules, and it seems that the identification and individuation of modules is exclusively post hoc.

Although none of these individual definitions will suffice as a theoretical account of modularity, the emphasis on interactivity in the definitions of Dover, Gilbert and Bolker, and Atchley and Hall is absolutely crucial to an adequate working definition. Too atomistic an account of modularity fails to take advantage of the truism that interactivity is the cheapest route to complexity, which helps to undergird the theorised role of developmental modularity in the evolution of evolvability (Kirschner and Gerhart 1998; Sterelny 2000b; Gilbert and Bolker 2001). Moreover, when we ask what practical end any account of modularity is supposed to serve, we find that internally integrated though highly interactive processes and entities provide an excellent framework for identifying and discussing collective and emergent properties, such as those emphasised throughout this book.[19]

One important consequence, then, is that at least some of the distinctive theses of DST are important in guiding biological practice and theory building, inasmuch as developmental systems theorists insist – even more so than evo–devoists or enlightened developmental biologists – that we focus on the constructive interactions generative of phenotypic complexity. So even though DST-style research programmes are already underway within evo–devo, it simply does not follow that DST is practically or theoretically irrelevant to the design of experiments or to how they are undertaken.

More directly relevant to establishing the practical importance of DST is the model of behavioural development described in Chapter 5 (Johnston and Edwards 2002), and also a model of developmental mechanisms recently postulated as the 'engine' of developmental systems theory. William Wimsatt has proposed that his model of *generative entrenchment* (GE) functions to mechanise key theses of DST, thereby adding substance to the verbal models expounded by DSTers (Wimsatt 2001: 219). Note, however, that Wimsatt's model of generative entrenchment was conceived independently of DST (Wimsatt 1986b, 1999; Schank and Wimsatt 1986, 2001; Wimsatt and Schank 1988). The model has obvious relevance to the emphasis within evo–devo on developmental constraints, modularity, and evolvability, but it also clearly relates to DST.

In describing the model in the context of the developmental systems perspective, Wimsatt (2001) begins with Lewontin's account of Darwin's Principles. Any evolving system must '1. have descendants that differ in their properties (*variation*), 2. some of which are heritable (*heritable* variation), and 3. have varying causal tendencies to have descendants (heritable variation *in fitness*)' (Wimsatt 2001: 220, citing Lewontin 1970). To evolve, a system must meet each of these criteria simultaneously: they are the logical conditions for the occurrence of an evolutionary process.

However, Wimsatt argues, in order to meet these three conditions, an organism must already have met two others: they must be '4. structures which are generated over time so they have a developmental history (*generativity*)', and there must be '5. some elements that have larger or more pervasive effects than others in that production (*differential entrenchment*)' (Wimsatt 2001: 221). Although these are not logical conditions for development in the way that Darwin's Principles are for evolution, Wimsatt knows of 'no interesting evolutionary process whatsoever (physical or conceptual) that does not meet them' (p. 220). Neither do I.

With these conditions stipulated, the generative entrenchment of an element of a structure is the magnitude of 'downstream' effects in an organism's life cycle borne by that element. That the magnitude of effects borne will vary between different elements makes GE a relative property (Wimsatt 2001: 221). Generatively entrenched elements are stable, persistent, and de facto foundational mechanisms involved in the generation of some number of ontogenetic effects. In fact, that there are recognisable life cycles at all, Wimsatt notes, is a consequence of GE – that is, of condition 5 in the context of conditions 1 through 4.

So, a system that meets conditions 1 through 3 does so by way of causal structures meeting the fourth and fifth conditions:

> They will thus have a development, and if they can reproduce and pass on their set of generators, will have a heredity. This order is expository, not causal: without a minimally reliable heredity, they cannot evolve a complex developmental phenotype, but developmental architecture can increase the efficacy and reliability of hereditary transmission. Heredity and development thus bootstrap each other, as emerging genotype and phenotype, through evolution. (Wimsatt 2001: 223)[20]

Thus, generatively entrenched elements are inherited, but they are not as a result necessarily gene-like. Those biological things marked as generatively entrenched might include genes but would also include non-genetic entities and, much more often, heterogeneous complexes of genetic and non-genetic entities (Wimsatt 2001: 224).

Several predictions follow from this model of GE: the one most relevant for our purposes is that changes in generators would have a higher probability of being (severely) maladaptive for the organism than changes in less deeply generatively entrenched elements, leading to evolutionary conservativism.[21] Deeply generatively entrenched elements are acutely embedded in the developing system, and they are burdened by sometimes huge numbers of developmental interactions. Such generators would therefore appear to be entirely unsuitable targets for natural selection.

The problem this raises is that we all know that Haeckelian and von Baerian notions of evolutionary change based on exclusively terminal additions are wrong. Moreover, we have learned from evo–devo that substantial reorganisations of development are possible without major alterations – or even *any* alterations – in adult phenotype (Raff 1996; Hall 1999; Dover 2000; Weiss and Fullerton 2000). So, how can GE be rendered compatible with the widely accepted idea that even developmentally precocious modular processes and entities are subject to change quasi-independently of each other? If GE is the engine of developmental systems theory, it appears that DST may well stall at the curb of evo–devo.

Wimsatt and Jeffrey Schank have recently responded to this problem. In their definition, modularity is

> a form of quasi-independence modulating the mappings between genotypes and phenotypes, allowing characters to vary in ways more likely to satisfy continuity and thereby increasing the proportion of phenotypic characters that may be adaptive. Thus, we analyze modularity in terms of the pleiotropic interactions of genes mapping to phenotypes. (Schank and Wimsatt 2001: 324)[22,23]

Schank and Wimsatt hold that at least some aspects of development must be modular – in their sense – in order that the variation generated by a change in development meet the requirement of continuity (Schank and Wimsatt 2001: 325). They also hold that GE is not in conflict with modularity, though generatively entrenched elements will be less modular than non-generatively entrenched or less generatively entrenched elements (p. 326).[24] The compatibility arises from the prediction that any complex adaptation will be the joint product of some modules and of more generatively entrenched elements (p. 327). (They suggest that this prediction appears to hold in at least some cases, such as eye development – no small accomplishment.) However, GE may well nonetheless be incompatible with the 'deep developmental modifications' generated by, say, heterochrony. So another, related prediction is that, given GE, heterochrony should be very rare, unless other mechanisms compensate in satisfying the requirements of quasi-independence and continuity (p. 327). In the case of both predictions, GE limits modularity in the generation of complex adaptions, but plausibly so.

If it is fair to suggest that GE as a model is compatible with the distinctive theses of developmental systems theory (and Wimsatt would so suggest), then, as this discussion of GE in relation to modularity has shown, it is also fair to suggest that a fruitful relationship between DST and evo–devo might be fostered through the elaboration of more DST-style models of developmental and evolutionary mechanisms.

THE THEORETICAL VALUE OF DST

Notwithstanding potential ways in which it may have an impact on experimental design, at the very least the developmental systems perspective guides us in our interpretation of experimental results. This is the third way in which developmental systems theory could be important to the establishment of a new biological synthesis. Evo–devoists are no less guilty than other biologists of illicit inferences on the basis of the data at hand. As the epistemological limits of our inferences are circumscribed by the limitations of our methodologies, and until methodological systemism is more widely appreciated, the developmental systems perspective as a philosophy of science is useful in helping to guard against invalidly inferred conclusions (of the sort explored in Chapter 1, for instance).

Nowhere is this need more evident than in the domain of public pronouncements made on behalf of genetics and genomics research. It is rather difficult not to be swept up in the hype and hoopla surrounding the publication of the human genome sequence. News magazines, scientific journals, and press conferences have served as soapboxes for predictions of genomics' potential to alleviate physical and social problems – from cancer to schizophrenia, from aggression to criminality. Of course, no one seriously thinks that the genome will have an immediate impact on human health or behaviour, for there are pesky developmental interactions aplenty which intervene between genomes and individual phenotypes. Such relationships must be elucidated in detail before we can begin to make sense of the mass of data contained in human genome sequence databases.

Even so, hopes are high, genomic dreams persist, and very many of us have been led to believe that comfort and security are to be found in the sequenced nucleotide bases of DNA. Hence the enormous pots of public funds made available to researchers in genetics and genomics, and the relatively paltry amounts procurable by those investigating higher-level determinants of health and disease (Strohman 1997; Bains 2001). However, we must be ever vigilant in assessing the inferences drawn from genomic sequence data, especially in light of the discussion of heuristics in Chapter 1, else we get swept up in the enchanting dream that the post-genomic era will be rife with new therapies for all that ails us. To this end, the distinctive theses of developmental systems theory make a significant contribution.

TAKING DEVELOPMENT SERIOUSLY

What, then, of the significance of developmental systems theory? That is, how apt is Sterelny's assertion that DST is irrelevant to the synthesis of evolutionary and developmental biology? First, let us consider some limitations of evolutionary developmental biology in light of DST. If evo–devo aims to explain how development impinges on evolution, and how development itself evolves, from the genocentric stance it occasionally adopts, then evo–devo will be merely a shadow of what it could be. If evo–devo interprets epigenetic events as nothing but the regulation of expression of genetic information, then it will provide no explanation of *development* as such, but only of *gene activation in development*, which is but a subsidiary question. Of course, if we define development in accordance with the modern consensus as differential gene expression, then a genocentric focus just is a developmental focus; but I see no persuasive reason to define development in such limited – and limiting – terms.

In contrast, my accounts of constitutive epigenetics and the creativity of development force a broad interpretation of evo–devo, one according to which the development of whole organisms and the evolution of their modular parts are deemed the primary analysands, rather than as secondary to the epigenetic expression of purely or primarily genetic potential. Such is the force of taking development seriously.

A developing system is clearly organised, but in a systemic way; that is, the interrelations between its parts are structured into causal, generative systems with ontogenetic control dispersed throughout these systems. Moreover, organisms are more than epiphenomena of genomes, more even than epiphenomena of genomes in particular structured environments. For a genome is in no sense prior to or separate from an organism, and an organism is in no sense prior to or separate from an environment. The organism is part of the environment (and the environment part of the organism), whereas the genotype is part of the phenotype; in fact, not only does an organism require an environment, so too does a genome prerequire an organism for its very expression.

However, we need not presume that organisms (and their development) cannot be explained at all – they cannot be explained atomistically, perhaps, but they can indeed be explained systemically by appeal to constitutive relations between components of their composition, structure, and environment. Yet, depending on our explanatory goals, sometimes focusing on organisms may not be enough; this is the pragmatic dimension of explanation (Gannett 1999; van der Weele 1999), and so we must accept the need for multiple units at the core of any adequate synthesis of evolutionary and developmental biology.

Evolutionary developmental biology has in its best incarnations the virtue of considering 'organisms as more than adults, embryos as more than means of making adults, and the phenotype as more than the physical expression of the genotype' (Hall 1999: 399). Evo–devo would do well, moreover, to underscore with DST that inheritance does not reduce to genetic transmission – hereditary potential is emergent, and a function of the developmental manifold, not merely the genome (Jablonka and Lamb 2002).

Of course, developmental systems theory is itself limited and has much to learn from evo–devo, as well. For instance, as yet, DST has had too little to say about a number of issues well addressed within evolutionary developmental biology, including the evolution of developmental mechanisms; the nature of developmental integration and developmental modularity, and their interrelations; the evolution of evolvability; the source of macroevolutionary episodes; the idea that individual ontogenies are the source of potentially evolutionarily significant variation at the phenotypic level; and the concepts of homology and homoplasy and their proper roles in developmental and evolutionary biology. These are serious lacunae in developmental systems theory. Coupled with the merely promissory verbal models of developmental processes and their role in evolution (with the possible exception of Wimsatt's model of generative entrenchment, and Johnston and Edwards' model of behavioural development explored in Chapter 5), developmental systems theorists have much substantial work left ahead of them. In sum, then, evolutionary developmental biologists and developmental systems theorists would do well to interact with each other in establishing a genuinely synthetic biology.

Let me reiterate the overarching theme of this book that no one denies that complex interactions are at the heart of the analysis of organismal development, both ordered and disordered. We can all agree that development is a matter of contingent interactive processes between multiple components within hierarchical systems in specific (though variable) contexts, and these must be investigated in depth if we are ever to provide an adequate account of ontogeny and of its relation to evolution. The details, then, the minutiae of what remains to be discovered, are not what's at issue in the dispute over the nature of a genuinely synthetic biology. Why, then, have the limitations of genes-*plus* accounts of interaction not been more widely recognised and appreciated? Why do modern consensus metaphors of genetic programs and primacy persist? Why are organisms still so often portrayed as basically or ultimately the product of genes?

If not the gory details, what *is* in dispute is *taking seriously* – and so putting into practice – what we all so easily grant in theory. The Devil is not in the details, but rather in the Gestalt. It's high time he be exorcised.

Endnotes

1. The phrase 'taking development seriously' is used independently by several authors whose views are in many ways compatible with my own, for example, as a section heading by Griffiths and Gray (1994) and by Oyama (2000a: 110), who writes that 'resolving the nature–nurture dichotomy requires, ironically, taking development seriously'.

CHAPTER I

1. For an extended discussion of hedgeless hedging (though not referred to by that epithet) in the case of genetic research on the aetiology of schizophrenia, see Robert (2000a).
2. I include 'predict', 'compute', and 'construct', as Rosenberg and Wolpert both slide casually from one to the next.
3. The claim that a given property could not have been explained is stronger than that it could not have been predicted, inasmuch as we can explain much that we could not have predicted; see, for instance, Gilbert and Sarkar (2000: 3).
4. In a review of loss-of-function experiments involving homeobox genes in the mouse skeleton, Smith and Schneider (1998) are highly critical of a number of studies in which such an illicit inference is drawn, usually as part of a more general claim of the evolutionary role of the homeobox genes in producing skeletal novelties. See also Robert (2001a).
5. This embedded reference is to W. George, *Elementary Genetics*, 2nd ed. (London: Macmillan, 1964), p. 46.
6. Emergent properties can be contrasted with resultant properties: if some x has a property p possessed as well by at least one of the components of x, then p is said to be resultant; but if x has some property q not to be found amongst the components of x, then q is said to be an emergent property of x. The property 'being alive' is a resultant property of multicellular organisms, but an emergent property of cells (Mahner and Bunge 1997: 29). This account of emergence is thus ontological, not epistemological. Emergence is not illusory – it will not disappear once we understand the relevant processes more completely – but is rather an inescapable feature

131

of the world. (Nevertheless, emergence may represent a problem for epistemology as well, for emergent properties cannot always be explained and predicted from our knowledge of lower-level systems.) For recent philosophical treatments of emergence, see, for example, Spencer-Smith (1994–1995), Humphries (1996), Wimsatt (1997; also see his 1986a), and Kim (1999).

7.   If the emergentist-materialist ontology underlying biology (and, as a matter of fact, all the factual sciences) is correct, the *bios* constitutes a distinct ontic level the entities in which are characterized by emergent properties. The properties of biotic systems are then not (ontologically) reducible to the properties of their components, although we may be able to partially explain and predict them from the properties of their components ... The belief that one has reduced a system by exhibiting [for instance] its composition, which is indeed nothing but physical and chemical, is insufficient: physics and chemistry do not account for the structure, in particular the organization, of biosystems and their emergent properties (Mahner and Bunge 1997: 197).

CHAPTER 2

1.   As Maienschein (1991b) has shown, the work of Wilhelm His (1834–1901) on germinal localisation was an important backdrop; also see Moore (1993).

2.   'For example, in the few cases in frogs in which a whole embryo did result from the one blastomere, he suggested that there exists a reserve idioplasm (or set of nuclear materials). This reserve comes into action in the special cases when regeneration or postgeneration (following injury) occurs' (Maienschein 1991b: 51).

3.   In 1910, J.F. McClendon repeated Roux's experiments, but, having separated the blastomeres, he obtained Driesch's results; see McClendon (1910) and Gilbert (2000a: 61).

4.   Maienschein (1991b: 51), citing Hans Driesch, 'Entwicklungsmechanische Studien I. Der Werth der beiden ersten Furchungszellen in der Echinodermentwicklung. Experimentelle Erzeugen von Theil – und Doppelbildung', *Zeitschrift für wissenschaftliche Zoologie* 53 (1891): 160–178, translated and abridged in B. Willier and J.M. Oppenheimer (eds.), *Foundations of Experimental Embryology* (New York: Hafner Press, 1974), pp. 38–50, at p. 46 of the translation. The full passage, evincing Driesch's astonishment, is as follows: 'I must confess that the idea of a free-swimming hemisphere or a half gastrula with its archenteron open lengthwise seemed rather extraordinary. I thought the formations would probably die. Instead, the next morning I found in their respective dishes typical, actively swimming blastulae of half size.'

5.   I thank Brian Hall for bringing this case to my attention. Hall thinks that the results are provocative, though he is careful not to read too much into findings regarding only a single species. See his remarks in Vogel (2000: 2120).

CHAPTER 3

1.   Needham (1959: 40) remarks that the antithesis between epigenesis and preformation is coextensive with the history of embryology. For the pre-twentieth-century history of the 'antithesis' see, for example, Needham's seminal work (1959) and

also Pinto-Correia's entertaining tale from the perspective of the vanquished – that is, of the preformationists (1997). As we shall see, preformationists have arisen anew in modern garb, and with a host of new problems in addition to some of the old ones.

2. This position may be rendered more or less palatable by specifying what are to count as constituents; but my concern in what follows depends less on the content and more on the form of Monod's claim.

3. A pre-DNA-era perspective is that of E.B. Wilson, writing in 1925: 'In respect to a great number of characters, *heredity is effected by the transmission of a nuclear preformation which in the course of development finds expression in a process of cytoplasmic epigenesis*' (Wilson 1925: 1112; emphasis in the original). Note, however, that Mayr would appear to be making a still stronger point, namely that epigenesis is *directed by* the 'nuclear preformation'.

4. For a sample of the 'premodern consensus', as it were – apparent in the writings of C.O. Whitman, for instance – see Whitman (1894); for discussion, see Maienschein (1986), but beware of some uncharacteristic anachronisms in her presentation of his views. One example is this: Whitman asks 'how far is post-formation to be explained as the result of preformation, and how far as the result of external influences?' (Whitman 1894: 221) – a question that Maienschein renders as 'How much depends on the developmental response to external conditions rather than on *programmed* internal unfolding?' (Maienschein 1986: 91; my emphasis). It must surely be recognised that Whitman in his lecture during the summer of 1894 had no concept of an ontogenetic *programme*, for the use of computer language in biology is of much more recent origin. In placing words in Whitman's mouth, Maienschein evidences the propriety of Keller's argument that with the rise of processor-mediated language, computer-age concepts dominate our biological imagination (Keller 1995: 118) – even facilitating the misrepresentation of historical positions.

5. Of course, DNA, as an inert molecule, cannot replicate itself but must rather be replicated in the process of ontogeny.

6. In this passage, I have deleted references to Levins and Lewontin (1985) and Nijhout (1990) at the end of the first sentence, and to Lewin (1997) after the word 'translation' in the second sentence. As will become evident in later chapters, I would expand consideration beyond the space between DNA and the phenotype, focusing instead on that between the egg and the mature organism.

7. Another possible epithet is *constructive development*, but the word 'constructive', even when intended literally – and not in the sense implied by social constructionists (whatever that may be: see, e.g., Hacking 1999) – nonetheless raises hackles. Of course, so too might the word 'creative', especially in the light of Bergson's 'creative evolution' – but I am more comfortable with the latter association than with the former. Gilbert Gottlieb has elaborated a position he calls 'probabilistic epigenesis' (reviewed in Gottlieb 1998, for instance), which bears some affinity with my own view; Bidell and Fischer (1997) rename Gottlieb's position 'constructive epigenesis' – a further reason to offer a distinct euphemism for my distinct view.

8. The concept of the 'integron' is borrowed from François Jacob: 'At each level, units of relatively well-defined size and almost identical structure associate to form a unit of the level above. Each of these units formed by the integration of sub-units may be given the general name 'integron'. An integron is formed by assembling

integrons of the level below it; it takes part in the construction of the level above' (Jacob 1973, as cited by Mayr 1997: 19).

9. This latter claim is particularly surprising in that Maynard Smith himself modelled how non-genetic information could be inherited (so-called epigenetic inheritance) in his earlier work (1990).

10. Mahner and Bunge (1997: 281) suggest, somewhat rashly, perhaps, that 'information' has by now become an 'all-purpose term' in biology. 'It sounds very scientific, and seemingly indicates some deep insight, but it is often nothing but a disguise of ignorance, inviting people to proceed according to the rule "If you don't know what it is, call it *information*".'

11. The discussion here roughly follows that in Robert (2000d) and Robert (2001a).

12. Hence the euphemism 'developmental contructionism', another name for the developmental systems perspective. Compare Burian (1997: 259–260):

> It is clear that many incredibly intricate multi-level domains, mechanisms, processes, structures, and so on enter into development. Furthermore, many of these are formed (pardon the pun) 'on the fly' – that is, they are not laid out in advance but arise in interactions between genes and proteins that come to form a rapidly-shifting tartan of boundaries between domains and define something like morphological fields in the midst of ongoing processes of cell-type specification, tissue formation, organogenesis, etc. At any stage of development some of the relevant modules that enter into normal development preexist, others are formed in the course of events, and others will or will not be formed according to the status and condition of interacting units and modules at key moments in the processes in question.

13. The notion of a genetic blueprint fares no better; see Mahner and Bunge (1997: 283), and particularly Oyama (1985) and Neumann-Held (1999).

14. The embedded quotations are from Mayr (1982); the same claims are reiterated in Mayr (1997).

CHAPTER 4

1. Cited in Gilbert (1988: 317, personal communication from Berrill to Gilbert, 1985).

2. Cited by Sander (1986: 368, translated by Sander from Hans Spemann, 'Vererbung und Entwicklungsmechanik', *Zeitschrift fur induktive Abstammungs und Vererbungslehre* 33 [1924]: 272–294, at p. 293).

3. Waddington was convinced that a complete account of ontogeny would include reference to genes, but only in the context of the developing embryo; the account would therefore emanate from a genetically informed embryology and not from genetics proper. For discussion, see, for example, Gilbert (1991a). Goodwin is a proponent of developmental structuralism, within which genes are largely irrelevant. Hull (1998) is an accessible review essay of Goodwin and Webster's *Form and Transformation* (Cambridge: Cambridge University Press, 1996). For a critique of developmental structuralism, see Smith (1993) and Mahner and Bunge (1997). Oyama (1985) is the *locus classicus* of developmental systems theory. Rose (1997) is a wholesale rejection of gene centrism.

4. The following four paragraphs are adapted from Robert (2000a: 198–199).
5. On epigenetics in its various formulations, see the collection edited by Russo et al. (1996) on epigenetic mechanisms of gene regulation, including the overview by Riggs and Porter (1996); also see Holliday's (1994) introduction to a special issue of *Developmental Genetics* (vol. 15, no. 6) devoted to epigenetics; Henikoff and Matzke's (1997) introduction to a special issue of *Trends in Genetics* (vol. 13, no. 8) on epigenetic effects; and Lewin's (1998) introduction to a special issue of *Cell* (vol. 93, no. 3) devoted to dispelling the 'mystique' of epigenetics. Also see Jablonka and Lamb's controversial book (Jablonka and Lamb 1995) on epigenetic inheritance, and especially the twin reviews by Griesemer (1998) and Keller (1998) in *Biology and Philosophy*, as well as the many reviews of their target article (Jablonka and Lamb 1998) in the *Journal of Evolutionary Biology* for 1998 (vol. 11, no. 2); both sets of commentaries include a response by the authors.
6. At the end of the first sentence quoted, Henikoff and Matzke refer to Holliday (1987).
7. Wolffe attributes this thesis to S.-Y. Lin and A.D. Riggs, 'The General Affinity of *lac* Repressor for *E. coli* DNA: Implication for Gene Regulation in Procaryotes and Eucaryotes', *Cell* 4 (1975): 107–111. Thus does Wolffe conclude that 'chromatin, chromosomes and nuclear structure itself are now known to be compartmentalized with respect to function' (1998: 2).
8. Keller cites Michael J. Apter, *Cybernetics and Development* (Oxford: Pergamon Press, 1966). It is noteworthy that Apter co-authored a 1965 paper with Lewis Wolpert (Apter and Wolpert 1965), in which they argued that developmental instructions are not localised at particular sites within the organism but rather that the system develops as a dynamic, integrated whole. Clearly this was written some time before Wolpert was converted to the "genetic programme" paradigm (which had occurred, according to Keller, by 1975). As will become subsequently evident, I am much more sympathetic toward this early view than toward Wolpert's later views (as in his 1991, 1994, and 1995).
9. It is rare to find mention of Russell in the biological or philosophical literature outside of discussions of comparative morphology, where his *Form and Function* (Russell 1916) is justly regarded as a classic. I am aware of only two modern treatments of Russell's *Interpretation* (1930) in particular: Nagel (1961) and Roll-Hansen (1984) – both of which are harshly (and, to my mind, uncharitably) critical.
10. For an extended critique of the ghost in the machine, see Oyama (1985).
11. For Rose, a 'lifeline' is an organism's 'unique trajectory through time and space' (1997: 98); also see Robert (2000d).
12. In this, I am following Neumann-Held (1999: 125) (also see Figure 6):

> The analysis of the molecular mechanisms of polypeptide expression shows quite clearly that there is no fundamental way by which the classical-molecular gene concept could be applied to DNA segments. One focuses at the same bit of DNA, and different structures and functions appear. One focuses on different levels of the expression process (DNA, primary mRNA, mature mRNA, edited mRNA, polypeptide), and again different structures

and functions appear. Introns can become exons, which can become pro-moters, and so on. Regarding the aspect of function, there is no general rule that a particular sequence codes for only one polypeptide. Also, in prin-ciple, no discrete material unit segment on the DNA can be identified as coding for (only) one polypeptide – at least not in the sense of the classical-molecular gene concept. Therefore, this gene concept is no longer useful; it is 'dead'.

13. Neumann-Held's 'developmental process gene' concept both defines genes as pro-cesses (which I do not) and also stops at the polypeptide chain; in extending the notion beyond the polypeptide chain, I am adopting the critique of Robert Perlman (personal communication, July 1999) that polypeptide chains do not actually exist but are merely convenient (and sometimes inconvenient) fictions.
14. I submit that this account of constitutive epigenetics is congruent with Newman and Müller's equally idiosyncratic 'pre-Mendelian' (or pre-genetic) interpretation of 'epigenetic mechanisms', introduced in Chapter 2.
15. The term 'mRNA' refers to messenger ribonucleic acid, which carries 'information' to ribosomes during transcription.

CHAPTER 5

1. For Keller's reasons for wishing to preserve the computer metaphor, see Keller (1999).
2. I owe the word and concept to Rose (1997: 18), who in turn owes them to Humberto Maturana.
3. Gillian Gass has observed (personal communication, July 2001) that these are both actually forms of inertia disturbed, respectively, by 'pushing' and by 'pulling'.
4. Though an exploration of the units of selection controversy is beyond the scope of this book, it is worth noting briefly that the perspective of genic selectionism is transformed in being forced to take niche construction into account. For now, from the point of view of genic selectionism, the constructive activities of organisms which translate into alterations in the sources of selection may eventually feed back into selection for the genes putatively responsible for particular niche-constructing traits (Odling-Smee et al. 1996: 643, 645–646).
5. The next three paragraphs closely follow Robert (2003).

CHAPTER 6

1. This chapter draws heavily on Robert (2002).
2. Sarkar (1998; Gilbert and Sarkar 2000) makes a helpful distinction between two kinds of reductionism – genetic reductionism and physical reductionism. Physi-cal reductionism sees physics as the most basic of the sciences and holds that all scientific explanations may (and should) eventually be recast in the terms of physics; genetic reductionism holds that 'genes can explain all phenotypic features of an organism' (Sarkar 1998: 174). These types of reductionism are not coextensive, though both champions and critics of reductionism tend to conflate these two va-rieties; but whereas physical reductionism is a thesis about relations between the

sciences, genetic reductionism is a thesis about the role of genes in organismal development. Both varieties of reductionism may be problematic, though for different reasons.

3. Both quotations are from Balfour and cited in Hall (2000b): the first, cited at p. 721, is from F.M. Balfour, 'A Preliminary Account of the Development of the Elasmobranch Fishes', *Quarterly Journal of the Microscopic Sciences* 4 (1874): 323–364, at p. 343; the second, cited at p. 722, is from F.M. Balfour, *A Treatise on Comparative Embryology*, 2 volumes (London: Macmillan & Co., 1880–1881), vol. 2, at p. 381.

4. For more on these problems of nomenclature, see Sarkar and Robert (2003), Gilbert (2003), and Love (2003).

5. Core topics of study in evo–devo not explicitly discussed here include the following: cell determination and differentiation; cell lineages; embryonic inductions; segmentation and compartmentalisation; heterochrony; homology and homoplasy; larval evolution (life cycle stages sometimes evolve independently); life history strategies; and inferences about the fossil record. For an overview, see Hall (1999). Hall and Olson (2003) is an encyclopaedia of keywords and concepts in evo–devo addressing these and other elements of the field.

6. Note that a *complete* catalogue of changes in ontogenetic pathways, adult phenotype, and gene frequency in a population is practically impossible for most lineages (possibly excepting fruit flies, bacteria, and viruses). The goal of a response to the charge of incompleteness then must not be (in the case of most lineages) completeness as such, but rather non-arbitrary representativeness; that is, proffering a representative catalogue of evolutionarily significant changes at multiple levels should be considered an appropriate, tractable response to the charge of incompleteness. Amundson (2001) offers a helpful critical analysis of the charge of completeness.

7. I borrow this example from Wagner (2000: 96–97), who discusses the work of Keys et al. (1999; see also Brakefield et al. 1996; Nijhout 1996; Brakefield and French 1999; and Brakefield 2001).

8. Experimental work on the causal role of the CR in the placement of the ribs was conducted by Burke (1989, 1991) – who also coined the term 'carapacial ridge'.

9. Note that palaeontologists have known this all along, in that palaeontologists have historically 'seen' only the phenotype. This helps to explain why the theory of punctuated equilibrium originated within palaeontology and not elsewhere. Thanks to Wendy Olson for this observation.

10. Sterelny has suggested (K. Sterelny, personal communication, May 2001) that Neo-Darwinians explore numerous non-externalist (though not obviously internalist) aspects of evolution, such as sexual selection and frequency-dependent selection. Pigliucci and Schlichting (1997: 147, 151) argue that quantitative geneticists' assumption of a constant fitness landscape ignores frequency-dependent selection. At any rate, the existence of such putative exceptions as Sterelny identifies makes Neo-Darwinism not *exclusively* externalist, though it may still be *excessively* externalist, in which case Arthur's complaint holds.

11. Here I draw on David Ingram's (2000: 86) discussion of cultural identity.

12. Resources for further consideration of the nature of disciplinary integration and synthesis can be found in Darden and Maull (1975), Bechtel (1986, 1993), and Robert (in preparation b).

CHAPTER 7

1. A fourth position, one I neglect here, is what is known as 'developmental structuralism', but see Smith (1993) and van der Weele (1999) for critical assessments of this position.

2. Both Szathmáry (1999) and Arthur (2000: 55) underscore that evidence in support of such a scenario just does not exist.

3. Another of Budd's examples is that of bithoracic flies (1999: 327), that is, flies with an extra pair of homeotically induced wings, who are nonetheless incapable of flying, and this for two reasons: the musculature of the fly is not accordingly altered homeotically to accommodate the extra wings and, more basically, the body plan of the fly is not aerodynamically suited to two pair of wings but only to the usual single pair. For further discussion, see Robert (2001a). See also Figure 3 in Chapter 2.

4. Perhaps Schwartz is an easy target, though; his expertise is in paleoanthropology, not developmental or evolutionary biology. However, then we need only recall the views explored in Chapters 3 and 4 to see that the modern consensus strikes a number of biologists as a perfectly adequate account of development, and so could serve to reconcile developmental and evolutionary biology under the aegis of genetics.

5. Hall therefore disagrees with those, such as Jablonka and Lamb (1995, 1998), who hold maternal cytoplasmic control to be an instance of epigenetic inheritance separate from genetic inheritance.

6. The developmental systems perspective is more often referred to as developmental systems theory, or DST. 'Theory', of course, is a term of art in the philosophy of science; because those who advocate a developmental systems perspective have generated no theory in the hypothetico-deductive sense of that term and have provided only verbal models in the model-theoretic sense of theory, finicky critics have suggested to me that 'perspective' may be more appropriate than 'theory'. (Note that this is not to prejudge the possible eventual elaboration of a developmental systems *theory* proper.) However, because developmental systems theory (and its acronym – DST) are in such widespread use, I shall use both epithets in what follows.

7. See, for example, Lehrman (1970). References to those whose work inspired the developmental systems perspective may be found in Johnston (1987), Gottlieb (1992, 1997), Oyama (1985, 2000a, 2000b), and Oyama et al. (2001).

8. See, for instance, work by Gottlieb (1992, 1995, 1997, 1998), Gray (1992, 2001), Griffiths and Gray (1994, 1997, 2001), and Oyama (1985, 2000a, 2000b). Also see Robert (2003).

9. In that article, Sterelny does not survey DST as a real live option; however, see Sterelny (2001), Sterelny and Griffiths (1999), and Sterelny et al. (1996) for more sustained coverage of the developmental systems perspective.

10. These include Sterelny et al. (1996), Schaffner (1998), and Godfrey-Smith (2000).

11. Wimsatt (2001: 229) cheekily proclaims that 'those who say that they don't need DST do not realize how much they have filched from it already or from common knowledge of developmental processes'.

12. Griffiths and Gray (1994) urge, in contrast, that developmental processes, not developmental systems, are central and foundational. Note that Gilbert and Bolker (2001), in their discussion of modularity and homology, underscore a similar need to focus on processes and changes in lieu of (strictly) structures.

13. Oyama et al. (2001) have no discrete section on the expanded set of interactants in development, though their discussion of the other theses leaves no doubt that they subscribe to this thesis as well.

14. Oyama et al. (2001: 3–4) use this same terminology.

15. Oyama, Griffiths, and Gray (2001) do not have a separate account of evolutionary developmental systems, though aspects of this thesis are discussed throughout their introductory essay.

16. For recent remarks on the developmental manifold, a term introduced almost thirty years ago by Gilbert Gottlieb, see Gottlieb (2002).

17. Griffiths and Neumann-Held (1999) distinguish between molecular and evolutionary gene concepts; for them, the evolutionary role that a so-called evolutionary gene must fill is that of 'heritable difference maker' (p. 661), and there is no reason that this role must, or should, be filled by a stretch of DNA.

18. Schaffner's surprise at this discovery is not unlike that of Scriver and Waters (1999) in uncovering the complexities of the causal pathways from gene to phene in the single-gene disorder phenylketonuria (PKU).

19. I thank Gillian Gass for gently nudging me toward these conclusions.

20. A 'generator' is an element with a relatively larger degree of GE (Wimsatt 2001: 221).

21. Here the affinity between GE and Von Baer's laws (briefly discussed in Chapter 6) becomes apparent: early embryonic structures, provided they are generatively entrenched, are far more difficult to change – they are evolutionarily and developmentally frozen – than later structures (even though not all early structures are generatively entrenched, and not all late structures are not).

22. 'Quasi-independence' is Lewontin's (1978) term for epigenetic pathways which interact only weakly with other such pathways between genes and characters.

23. 'Continuity' is Lewontin's (1978) term for the requirement that slight changes in a character lead to equally slight changes in the organism's relations with its environment – and thus to only small changes in reproductive fitness.

24. The discussion is actually of 'pleiotropic entrenchment', a special case of GE. The specific relationship between pleiotropic entrenchment and GE is irrelevant for our purposes here.

139

Bibliography

Akam, M. (1998). Hox genes, homeosis and the evolution of segment identity: No need for hopeless monsters. *International Journal of Developmental Biology*, **42**, 445–451.

Allen, G.E. (1978). *Life Science in the Twentieth Century*. Cambridge: Cambridge University Press.

(1986). T.H. Morgan and the split between embryology and genetics, 1910–35. In *A History of Embryology*, ed. T.J. Horder, J.A. Witkowski, and C.C. Wylie, pp. 113–146. Cambridge: Cambridge University Press.

Amundson, R. (1989). The trials and tribulations of selectionist explanations. In *Issues in Evolutionary Epistemology*, ed. K. Hahlweg and C.A. Hooker, pp. 413–432. Albany: State University of New York Press.

(1994). Two conceptions of constraint: Adaptationism and the challenge from developmental biology. *Philosophy of Science*, **61**, 556–578.

(2001). Adaptation and development: On the lack of common ground. In *Adaptationism and Optimality*, ed. S.H. Orzack and E. Sober, pp. 303–334. Cambridge: Cambridge University Press.

(5 January 2003). Changing the spin: An evo devo retelling of the birth of evolutionary biology. Paper presented at the Society for Integrative and Comparative Biology Annual Meeting, Toronto.

Apter, M.J., and Wolpert, L. (1965). Cybernetics and development, I: Information theory. *Journal of Theoretical Biology*, **8**, 244–257.

Aristotle (1953). *On the Generation of Animals*, trans. A.L. Peck. Cambridge, MA: Harvard University Press.

Arnold, S.J., Alberch, P., Csányi, V., Dawkins, R.C., Emerson, S.B., Fritzsch, B., Horder, T.J., Maynard Smith, J., Starck, M., Wagner, G.P., and Wake, D.B. (1989). How do complex organisms evolve? In *Complex Organismal Functions: Integration and Evolution in Vertebrates*, ed. D.B. Wake and G. Roth, pp. 403–433. Chichester: John Wiley & Sons.

Arthur, W. (1997). *The Origin of Animal Body Plans: A Study in Evolutionary Developmental Biology*. Cambridge: Cambridge University Press.

(2000). The concept of developmental reprogramming and the quest for an inclusive theory of evolutionary mechanisms. *Evolution & Development*, **2**, 49–57.

(2002). The emerging conceptual framework of evolutionary developmental biology. *Nature*, **415**, 757–764.

Atchley, W.R., and Hall, B.K. (1991). A model for development and evolution of complex morphological structures. *Biological Reviews of the Cambridge Philosophical Society*, **66**, 101–157.

Bains, W. (2001). The parts list of life. *Nature Biotechnology*, **19**, 401–402.

Bateson, W. (1894). *Materials for the Study of Variation Treated With Especial Regard to Discontinuity in the Origin of Species*. London: Macmillan & Co.

Bechtel, W. (1986). The nature of scientific integration. In *Integrating Scientific Disciplines*, ed. W. Bechtel, pp. 3–52. Dordrecht: Martinus Nijhoff.

(1993). Integrating disciplines by creating new disciplines: The case of cell biology. *Biology and Philosophy*, **8**, 277–329.

Berrill, N.J. (1941). Spatial and temporal growth patterns in colonial organisms. *Growth (Suppl.)*, **5**, 89–111.

Bidell, T.R., and Fischer, K.W. (1997). Between nature and nurture: The role of human agency in the epigenesis of intelligence. In *Intelligence, Heredity, and Environment*, ed. R.J. Sternberg and E. Grigorenko, pp. 193–242. Cambridge: Cambridge University Press.

Bolker, J.A., (1995). Model systems in developmental biology. *BioEssays*, **17**, 451–455.

(2000). Modularity in development and why it matters to evo-devo. *American Zoologist*, **40**, 770–776.

Bolker, J.A., and Raff, R.A. (1997). Beyond worms, flies and mice: It's time to widen the scope of developmental biology. *Journal of NIH Research*, **9**, 35–39.

Brakefield, P.M. (1998). The evolution-development interface and advances with the eyespot patterns of Bicyclus butterflies. *Heredity*, **80**, 265–272.

(2001). Structure of a character and the evolution of butterfly eyespot patterns. *Journal of Experimental Zoology (Molecular and Developmental Evolution)*, **291**, 93–104.

Brakefield, P.M., and French, V. (1999). Butterfly wings: The evolution and development of colour patterns. *BioEssays*, **21**, 391–401.

Brakefield, P.M., Gates, J., Keys, D., Kesbeke, F., Wijngaarden, P.J., Monteiro, A., French, V., and Carroll, S.B. (1996). Development, plasticity and evolution of butterfly eyespot patterns. *Nature*, **384**, 236–242.

Brakefield, P.M., and Kesbeke, F. (1997). Genotype-environment interactions for insect growth in constant and fluctuating temperature regimes. *Proceedings of the Royal Society of London B, Biological Sciences*, **264**, 717–723.

Brakefield, P.M., Kesbeke, F., and Koch, P.B. (1998). The regulation of phenotypic plasticity of eyespots in the butterfly *Bicyclus anynana*. *American Naturalist*, **152**, 853–860.

Brenner, S. (1974). New directions in molecular biology. *Nature*, **248**, 785–787.

Brylski, P., and Hall, B.K. (1988a). Ontogeny of a macroevolutionary phenotype: The external cheek pouches of geomyoid rodents. *Evolution*, **42**, 391–395.

(1988b). Epithelial behaviors and threshold effects in the development and evolution of internal and external cheek pouches in rodents. *Zeitschrift fur zoologische Systematik und Evolutionsforschung*, **26**, 144–154.

Budd, G. (1999). Does evolution in body patterning drive morphological change – or vice versa? *BioEssays*, **21**, 326–332.

Bibliography

Burian, R. (1997). On conflicts between genetic and developmental viewpoints – and their attempted resolution in molecular biology. In *Structures and Norms in Science*, ed. M.L. Dalla Chiara, K. Doets, D. Mundici, and J. van Benthem, pp. 243–264. Dordrecht: Kluwer.

Burke, A.C. (1989). Critical features in chelonian development: The ontogeny and phylogeny of a unique tetrapod bauplan. Unpublished PhD dissertation, Cambridge, MA: Harvard University.

(1991). The development and evolution of the turtle body plan: Inferring intrinsic aspects of the evolutionary process from experimental embryology. *American Zoologist*, **31**, 616–627.

Carroll, S.B., Grenier, J.K., and Weatherbee, S.D. (2001). *From DNA to Diversity: Molecular Genetics and the Evolution of Animal Design*. Malden, MA: Blackwell Science.

Chen, L., Krause, M., Draper, B., Weintraub, H., and Fire, A. (1992). Body-wall muscle formation in *Caenorhabditis elegans* embryos that lack the MyoD homolog *hlh-1*. *Science*, **256**, 240–243.

Danchin, A. (1996). On genomes and cosmologies. In *Integrative Approaches to Molecular Biology*, ed. J. Collado-Vides, B. Magasanik and T.F. Smith, pp. 91–111. Cambridge, MA: MIT Press.

Darden, L., and Maull, N. (1977). Interfield theories. *Philosophy of Science*, **44**, 43–64.

Darwin, C. (1859). *On the Origin of Species*. London: John Murray.

(1881). *The Formation of Vegetable Mold Through the Action of Worms With Observations on Their Habits*. London: John Murray.

von Dassow, G., and Munro, E. (1999). Modularity in animal development and evolution: Elements of a conceptual framework for evodevo. *Journal of Experimental Zoology (Molecular and Developmental Evolution)*, **285**, 307–325.

Dawkins, R. (1976). *The Selfish Gene*. Oxford: Oxford University Press.

(1986). *The Blind Watchmaker*. London: Longman.

Dennett, D.C. (1995). *Darwin's Dangerous Idea: Evolution and the Meanings of Life*. New York: Touchstone/Simon and Schuster.

Dover, G. (2000). How genomic and developmental dynamics affect evolutionary processes. *BioEssays*, **22**, 1153–1159.

Dunn, L.C. (1917). Nucleus and cytoplasm as vehicles of heredity. *American Naturalist*, **51**, 286–300.

Eldredge, N., and Gould, S.J. (1972). Punctuated equilibria: An alternative to phyletic gradualism. In *Models in Paleobiology*, ed. T.J.M. Schopf, pp. 82–115. San Francisco: Freeman.

Falk, R. (1991). The dominance of traits in genetic analysis. *Journal of the History of Biology*, **24**, 457–484.

Fraser, A. (1970). An epigenetic system. In *Towards a Theoretical Biology*, ed. C.H. Waddington, pp. 57–62. Chicago: Aldine Publishing Company.

Fusco, G. (2001). How many processes are responsible for phenotypic evolution? *Evolution & Development*, **3**, 279–286.

Gaertner, K. (1990). A third component causing random variability beside environment and genotype: A reason for the limited success of a 30 year long effort to standardize laboratory animals? *Laboratory Animals*, **24**, 71–77.

Bibliography

Gannett, L. (1999). What's in a cause? The pragmatic dimensions of genetic explanations. *Biology & Philosophy*, **14**, 349–374.

Garstang, W. (1922). The theory of recapitulation: A critical restatement of the Biogenetic Law. *Proceedings of the Linnean Society of London (Zoology)*, **35**, 81–101.

Gasser, S.M., Paro, R., Stewart, F., and Aasland, R. (1998). The genetics of epigenetics. *Cellular and Molecular Life Sciences*, **54**, 1–5.

Gerhart, J., and Kirschner, M. (1997). *Cells, Embryos and Evolution*. Malden, MA: Blackwell Science.

Gehring, W.J. (1985). The homeo box: A key to the understanding of development? *Cell*, **40**, 3–5.

(1998). *Master Control Genes in Development and Evolution: The Homeobox Story*. New Haven, CT: Yale University Press.

Gehring, W.J., Halder, G., and Callaerts, P. (1995). Induction of ectopic eyes by targeted expression of the *Eyeless* gene in *Drosophila*. *Science*, **267**, 1788–1792.

Gigerenzer, G., Todd, P.M., and the ABC Research Group. (1999). *Simple Heuristics That Make Us Smart*. Oxford: Oxford University Press.

Gilbert, S.F. (1988). Cellular politics: Just, Goldschmidt, and the attempts to reconcile embryology and genetics. In *The American Development of Biology*, ed. R. Rainger, K. Benson, and J. Maienschein, pp. 311–346. Philadelphia: University of Pennsylvania Press.

(1991a). Epigenetic landscaping: Waddington's use of cell fate bifurcation diagrams. *Biology & Philosophy*, **6**, 135–154.

(1991b). Induction and the origins of developmental genetics. In *A Conceptual History of Modern Embryology*, ed. S.F. Gilbert, pp. 181–206. Baltimore: Johns Hopkins University Press.

(1994). Dobzhansky, Waddington, and Schmalhausen: Embryology and the Modern Synthesis. In *The Evolution of Theodosius Dobzhansky: Essays on His Life and Thought in Russia and America*, ed. M.B. Adams, pp. 143–154. Princeton, NJ: Princeton University Press.

(2000a). *Developmental Biology*, 6th ed. Sunderland, MA: Sinauer Associates, Inc.

(2000b). Diachronic biology meets evo-devo: C.H. Waddington's approach to evolutionary developmental biology. *American Zoologist*, **40**, 729–737.

(2001). Ecological developmental biology: Developmental biology meets the real world. *Developmental Biology*, **233**, 1–12.

(2003). Evo-devo, devo-evo, and devgen-popgen. *Biology & Philosophy*, **18**, 347–352.

Gilbert, S.F., and Bolker, J.A. (2001). Homologies of process and modular elements of embryonic construction. *Journal of Experimental Zoology (Molecular and Developmental Evolution)*, **291**, 1–12.

Gilbert, S.F., and Faber, M. (1996). Looking at embryos: The visual and conceptual aesthetics of emerging form. In *The Elusive Synthesis: Aesthetics and Science*, ed. A.I. Tauber, pp. 125–151. Dordrecht: Kluwer.

Gilbert, S.F., and Jorgensen, E.M. (1998). Wormholes: A commentary on K.F. Schaffner's 'Genes, behavior, and developmental emergentism'. *Philosophy of Science*, **65**, 259–266.

Gilbert, S.F., Loredo, G.A., Brukman, A., and Burke, A.C. (2001). Morphogenesis of the turtle shell: The development of a novel structure in tetrapod evolution. *Evolution & Development*, **3**, 47–58.

Gilbert, S.F., Opitz, J.M., and Raff, R.A. (1996). Resynthesizing evolutionary and developmental biology. *Developmental Biology*, **173**, 357–372.

Gilbert, S.F., and Sarkar, S. (2000). Embracing complexity: Organicism for the 21st century. *Developmental Dynamics*, **219**, 1–9.

Godfrey-Smith, P. (2000). Explanatory symmetries, preformation, and developmental systems theory. *Philosophy of Science (Proceedings)*, **67**, S322–S331.

Goldsmith, H.H., Gottesman, I.I., and Lemery, K.S. (1997). Epigenetic approaches to developmental psychopathology. *Development & Psychopathology*, **9**, 365–387.

Gottlieb, G. (1971). *Development of Species Identification in Birds: An Inquiry into the Prenatal Determinants of Perception*. Chicago: University of Chicago Press.

(1992). *Individual Development and Evolution: The Genesis of Novel Behavior*. Oxford: Oxford University Press.

(1995). Some conceptual deficiencies in 'developmental' behavior genetics. *Human Development*, **38**, 131–141.

(1997). *Synthesizing Nature-Nurture: Prenatal Roots of Instinctive Behavior*. Mahwah, NJ: Lawrence Erlbaum Associates.

(1998). Normally occurring environmental and behavioral influences on gene activity: From Central Dogma to probabilistic epigenesis. *Psychological Review*, **105**, 792–802.

(2002). On the epigenetic evolution of species-specific perception: The developmental manifold concept. *Cognitive Development*, **17**, 1287–1300.

Gould, S.J. (1977). *Ontogeny and Phylogeny*. Cambridge, MA: Harvard University Press.

(2002). *The Structure of Evolutionary Theory*. Cambridge: Harvard University Press.

Gould, S.J., and Lewontin, R.C. (1979). The Spandrels of San Marco and the Panglossian Paradigm: A critique of the adaptationist programme. *Proceedings of the Royal Society of London B, Biological Sciences*, **205**, 581–598.

Gray, R. (1992). Death of the gene: Developmental systems strikes back. In *Trees of Life: Essays in the Philosophy of Biology*, ed. P.E. Griffiths, pp. 165–209. Dordrecht: Kluwer.

(2001). Selfish genes or developmental systems? In *Thinking About Evolution: Historical, Philosophical, and Political Perspectives*, ed. R.S. Singh, C.B. Krimbas, D.B. Paul, and J. Beatty, pp. 184–207. Cambridge: Cambridge University Press.

Griesemer, J. (1998). Turning back to go forward. *Biology & Philosophy*, **13**, 103–112.

(2000). Reproduction and the reduction of genetics. In *The Concept of the Gene in Development and Evolution*, ed. P.J. Beurton, R. Falk, and H.-J. Rheinberger, pp. 240–285. Cambridge: Cambridge University Press.

Griffiths, P.E., and Gray, R. (1994). Developmental systems and evolutionary explanation. *Journal of Philosophy*, **91**, 277–304.

(1997). Replicator 2 – Judgement Day. *Biology & Philosophy*, **12**, 471–492.

(2001). Darwinism and developmental systems. In *Cycles of Contingency: Developmental Systems and Evolution*, ed. S. Oyama, P.E. Griffiths, and R. Gray, pp. 195–218. Cambridge, MA: MIT Press.

Griffiths, P.E., and Knight, R.D. (1998). What is the developmentalist challenge? *Philosophy of Science*, **65**, 253–258.

Griffiths, P.E., and Neumann-Held, E.M. (1999). The many faces of the gene. *BioScience*, **49**, 656–674.

145

Hacking, I. (1999). *The Social Construction of What?* Cambridge, MA: Harvard University Press.

Hall, B.K. (1992a). Waddington's legacy in development and evolution. *American Zoologist*, **32**, 113–122.

(1992b). *Evolutionary Developmental Biology*. London: Chapman and Hall.

(1998). Epigenetics: Regulation not replication. *Journal of Evolutionary Biology*, **11**, 201–205.

(1999). *Evolutionary Developmental Biology*, 2nd ed. Dordrecht: Kluwer.

(2000a). Evo-devo or devo-evo – Does it matter? *Evolution & Development*, **2**, 177–178.

(2000b). Balfour, Garstang and de Beer: The first century of evolutionary embryology. *American Zoologist*, **40**, 718–728.

(2003). Opening the black box between genotype and phenotype: Cells and cell condensations as fundamental units of evolutionary developmental biology. *Biology & Philosophy*, **18**, 219–247.

Hall, B.K., and Miyake, T. (1992). The membranous skeleton: The role of cell condensations in vertebrate skeletogenesis. *Anatomy and Embryology*, **186**, 107–124.

(1995). Divide, accumulate, differentiate: Cell condensation in skeletal development revisited. *International Journal of Developmental Biology*, **39**, 881–893.

(2000). All for one and one for all: Condensations and the initiation of skeletal development. *BioEssays*, **22**, 138–147.

Hall, B.K., and Olson, W.M., eds. (2003). *Keywords and Concepts in Evolutionary Developmental Biology*. Cambridge, MA: Harvard University Press.

Hamburger, V. (1980). Embryology and the Modern Synthesis in evolutionary theory. In *The Evolutionary Synthesis: Perspectives on the Unification of Biology*, ed. E. Mayr and W.B. Provine, pp. 97–112. Cambridge, MA: Harvard University Press.

Hamer, D. (2002). Rethinking behavior genetics. *Science*, **298**, 71–72.

Hamer, D., and Copeland, P. (1998). *Living with Our Genes: The Groundbreaking Book about the Science of Personality, Behavior, and Genetic Destiny*. New York: Anchor/Doubleday.

Harrison, R.G. (1918). Experiments on the development of the fore limb of Amblystoma, a self-differentiating equipotential system. *Journal of Experimental Zoology*, **25**, 413–461.

(1937). Embryology and its relations. *Science*, **85**, 369–374.

Henikoff, S., and Matzke, M.A. (1997). Exploring and explaining epigenetic effects. *Trends in Genetics*, **13**, 293–295.

van Heyningen, V. (2000). Gene games of the future. *Nature*, **408**, 769–771.

Hoagland, S.L. (1988). *Lesbian Ethics*. Palo Alto, CA: Institute of Lesbian Studies.

Holliday, R. (1987). The inheritance of epigenetic defects. *Science*, **238**, 163–170.

(1994). Epigenetics: An overview. *Developmental Genetics*, **15**, 453–457.

Hull, D.L. (1998). A clash of paradigms or the sound of one hand clapping. *Biology & Philosophy*, **13**, 587–595.

Humphreys, P. (1996). Aspects of emergence. *Philosophical Topics*, **24**, 53–70.

Ingram, D. (2000). *Group Rights: Reconciling Equality and Difference*. Lawrence: University Press of Kansas.

International Human Genome Sequencing Consortium. (2001). Initial sequencing and analysis of the human genome. *Nature*, **409**, 860–921.

Jablonka, E., and Lamb, M.J. (1995). *Epigenetic Inheritance and Evolution: The Lamarckian Dimension*. Oxford: Oxford University Press.

　(1998). Epigenetic inheritance in evolution. *Journal of Evolutionary Biology*, **11**, 159–183.

　(2002). Creating bridges or rifts? Developmental systems theory and evolutionary developmental biology. *BioEssays*, **24**, 290–291.

Jacob, F. (1973). *The Logic of Life: A History of Heredity*, trans. B.E. Spillmann. Princeton, NJ: Princeton University Press.

Jeffery, W.R. (2001). Cavefish as a model system in evolutionary developmental biology. *Developmental Biology*, **231**, 1–12.

Johnston, T.D., (1987). The persistence of dichotomies in the study of behavioral development. *Developmental Review*, **7**, 149–182.

Johnston, T.D., and Edwards, L. (2002). Genes, interactions, and the development of behaviour. *Psychological Review*, **109**, 26–34.

Johnston, T.D., and Gottlieb, G. (1990). Neophenogenesis: A developmental theory of phenotypic evolution. *Journal of Theoretical Biology*, **147**, 471–495.

Keller, E.F. (1994). Master molecules. In *Are Genes Us? The Social Consequences of the New Genetics*, ed. C.F. Cranor, pp. 89–98. New Brunswick, NJ: Rutgers University Press.

　(1995). *Refiguring Life: Metaphors of Twentieth-Century Biology*. New York: Columbia University Press.

　(1998). Structures of heredity. *Biology & Philosophy*, **13**, 113–118.

　(1999). Understanding development. *Biology & Philosophy*, **14**, 321–330.

　(2000). *The Century of the Gene*. Cambridge, MA: Harvard University Press.

　(2001). Beyond the gene but beneath the skin. In *Cycles of Contingency: Developmental Systems and Evolution*, ed. S. Oyama, P.E. Griffiths, and R. Gray, pp. 299–312. Cambridge, MA: MIT Press.

　(2002). *Making Sense of Life: Explaining Biological Development with Models, Metaphors, and Machines*. Cambridge, MA: Harvard University Press.

Keller, L., and Ross, K.G. (1993). Phenotypic plasticity and 'cultural transmission' of alternative social organisations in the fire ant *Solenopsis invicta*. *Behavioural Ecology & Sociobiology*, **33**, 121–129.

Keys, D.N., Lewis, D.L., Selegue, J.E., Pearson, B.J., Goodrich, L.V., Johnson, R.L., Gates, J., Scott, M.P., and Carroll, S.B. (1999). Recruitment of a hedgehog regulatory circuit in butterfly eyespot evolution. *Science*, **283**, 532–534.

Kim, J. (1999). Making sense of emergence. *Philosophical Studies*, **95**, 3–36.

Kirschner, M., and Gerhart, J. (1998). Evolvability. *Proceedings of the National Academy of Sciences U.S.A.*, **95**, 8420–8427.

Kitcher, P. (2001). Battling the undead: How (and how not) to resist genetic determinism. In *Thinking About Evolution: Historical, Philosophical, and Political Perspectives*, ed. R.S. Singh, C.B. Krimbas, D.B. Paul, and J. Beatty, pp. 396–414. Cambridge: Cambridge University Press.

Krimsky, S. (2000). *Hormonal Chaos: The Scientific and Social Origins of the Environmental Endocrine Hypothesis*. Baltimore: Johns Hopkins University Press.

Laland, K.N.,Odling-Smee, F.J., and Feldman, M.W. (1999). Evolutionary consequences of niche construction and their implications for ecology. *Proceedings of the National Academy of Sciences U.S.A.*, **96**, 10242–10247.

(2001). Niche construction, ecological inheritance, and cycles of contingency in evolution. In *Cycles of Contingency: Developmental Systems and Evolution*, ed. S. Oyama, P.E. Griffiths, and R. Gray, pp. 117–126. Cambridge, MA: MIT Press.

Laubichler, M.D., and Wagner, G.P. (2001). How molecular is molecular developmental biology? A reply to Alex Rosenberg's 'Reductionism redux: Computing the embryo'. *Biology & Philosophy*, **16**, 53–68.

Lehrman, D.S. (1965). Interaction between internal and external environments in the regulation of the reproductive cycle of the ring dove. In *Sex and Behavior*, ed. F.A. Beach, pp. 335–369, 378–380. New York: John Wiley & Sons.

(1970). Semantic and conceptual issues in the nature-nurture problem. In *Development and Evolution of Behavior*, ed. L.R. Aronson, D.S. Lehrman, E. Tobach, and J.S. Rosenblatt, pp. 17–52. San Francisco: Freeman.

Levins, R., and Lewontin, R.C. (1985). *The Dialectical Biologist*. Cambridge, MA: Harvard University Press.

Lewin, B. (1997). *Genes VI*. New York: Oxford University Press.

(1998). The mystique of epigenetics. *Cell*, **93**, 301–303.

Lewis, E.B. (1978). A gene complex controlling segmentation in Drosophila. *Nature*, **276**, 565–570.

(1994). Homeosis: The first 100 years. *Trends in Genetics*, **10**, 341–343.

Lewontin, R.C. (1970). The units of selection. *Annual Review of Ecology and Systematics*, **1**, 1–18.

(1974). The analysis of variance and the analysis of causes. *American Journal of Human Genetics*, **26**, 400–411.

(1978). Adaptation. *Scientific American*, **239**, 156–169.

(1983). Gene, organism, and environment. In *Evolution from Molecules to Men*, ed. D.S. Bendall, pp. 273–285. Cambridge: Cambridge University Press.

Lickliter, R. (2000). An ecological approach to behavioral development: Insights from comparative psychology. *Ecological Psychology*, **12**, 319–334.

Lloyd, E.A. (1999). Evolutionary psychology: The burden of proof. *Biology & Philosophy*, **14**, 211–233.

Love, A.C. (2003). Evolutionary morphology, innovation, and the synthesis of evolutionary and developmental biology. *Biology & Philosophy*, **18**, 309–345.

Løvtrup, S. (1974). *Epigenetics: A Treatise on Theoretical Biology*. Toronto: John Wiley & Sons.

(1988). Epigenetics. In *Ontogeny and Systematics*, ed. C.J. Humphries, pp. 189–227. New York: Columbia University Press.

Mahner, M. and Bunge, M. (1997). *Foundations of Biophilosophy*. Berlin: Springer-Verlag.

Maienschein, J. (1986). Preformation or new formation – or neither or both? In *A History of Embryology*, ed. T.J. Horder, J.A. Witkowski, and C.C. Wylie, pp. 73–108. Cambridge: Cambridge University Press.

(1991a). *Transforming Traditions in American Biology, 1880–1915*. Baltimore: Johns Hopkins University Press.

(1991b). The origins of *Entwicklungsmechanik*. In *A Conceptual History of Embryology*, ed. S.F. Gilbert, pp. 43–61. New York: Plenum Press.

Marshall, E. (2000). Rival genome sequencers celebrate a milestone together. *Science*, **288**, 2294–2295.

Maynard Smith, J. (1990). Models of a dual inheritance system. *Journal of Theoretical Biology*, **143**, 41–53.

(2000a). The concept of information in biology. *Philosophy of Science*, **67**, 177–194.

(2000b). Reply to commentaries. *Philosophy of Science*, **67**, 214–218.

Mayr, E. (1960). The emergence of evolutionary novelties. In *Evolution After Darwin, Volume 1: The Evolution of Life, its Origin, History and Future*, ed. S. Tax, pp. 349–380. Chicago: University of Chicago Press.

(1982). *The Growth of Biological Thought: Diversity, Evolution, and Inheritance*. Cambridge, MA: Harvard University Press.

(1997). *This is Biology: The Science of the Living World*. Cambridge, MA: Harvard University Press.

Mazzeo, J.A., ed. (1977). Introduction. In O. Hertwig, *The Biological Problem of Today: Preformation or Epigenesis? The Basis of a Theory of Organic Development*. Oceanside, NJ: Dabor Science Publications.

McCain, R.A. (1980). Critical reflections on sociobiology. *Review of Social Economy*, **38**, 123–139.

McClendon, J.F. (1910). The development of isolated blastomeres of the frog's egg. *American Journal of Anatomy*, **10**, 425–430.

McGinnis, W., Levine, M.S., Hafen, E., Kuroiwa, A., and Gehring, W.J. (1984). A conserved DNA sequence in homeotic genes of the *Drosophila antennapedia* and *Bithorax* complex. *Nature*, **308**, 428–433.

Medawar, P.S., and Medawar, J.S. (1977). *The Life Science*. New York: Harper and Row.

(1983). *From Aristotle to Zoos: A Philosophical Dictionary of Biology*. Cambridge, MA: Harvard University Press.

Molenaar, P.C.M., Boomsma, D.I., and Dolan, C.V. (1993). A third source of developmental differences. *Behavior Genetics*, **23**, 519–524.

Monod, J. (1971). *Chance and Necessity*, trans. A. Wainhouse. New York: Knopf.

Moore, J.A. (1993). *Science as a Way of Knowing: The Foundations of Modern Biology*. Cambridge, MA: Harvard University Press.

Morgan, T.H. (1907). Sex-determining factors in animals. *Science*, **25**, 382–384.

(1909). What are 'factors' in Mendelian explanations? *American Breeders' Association*, **5**, 365–368.

(1919). *The Physical Basis of Heredity*. Philadelphia: J.B. Lippincott Co.

(1926). *The Theory of the Gene*. New Haven, CT: Yale University Press.

(1932a). The rise of genetics, I. *Science*, **76**, 261–267.

(1932b). The rise of genetics, II. *Science*, **76**, 285–288.

(1934). *Embryology and Genetics*. New York: Columbia University Press.

Moss, L. (1992). A kernel of truth? On the reality of the genetic program. *PSA 1992: Philosophy of Science Association (Proceedings)*, vol. 1, pp. 335–348.

Moss, M.L. (1981). Genetics, epigenetics, and causation. *American Journal of Orthodontics*, **80**, 366–375.

Müller, W.A. (1996). From the Aristotelian soul to genetic and epigenetic information: The evolution of the modern concepts in developmental biology at the turn of the century. *International Journal of Developmental Biology*, **40**, 21–26.

Nagel, E. (1961). *The Structure of Science: Problems in the Logic of Scientific Explanation*. New York: Harcourt, Brace & World.

Needham, J. (1959). *A History of Embryology*, 2nd ed. New York: Abelard-Schuman.

Neumann-Held, E.M. (1999). The gene is dead – Long live the gene! Conceptualizing genes the constructionist way. In *Sociobiology and Bioeconomics: The Theory of Evolution in Biological and Economic Theory*, ed. P. Koslowski, pp. 105–137. Berlin: Springer-Verlag.

Newman, S.A., and Müller, G.B. (2000). Epigenetic mechanisms of character origination. *Journal of Experimental Zoology (Molecular and Developmental Evolution)*, **288**, 304–317.

Nijhout, H.F. (1990). Metaphors and the role of genes in development. *BioEssays*, **12**, 441–446.

(1991). *The Development and Evolution of Butterfly Wing Patterns*. Washington, DC: Smithsonian Institution Press.

(1996). Focus on butterfly eyespot development. *Nature*, **384**, 209–210.

Odling-Smee, F.J., Laland, K.N., and Feldman, M.W. (1996). Niche construction. *American Naturalist*, **147**, 641–648.

van Oosterhout, C., and Brakefield, P.M. (1999). Quantitative genetic variation in *Bicyclus anynana* metapopulation. *Netherlands Journal of Zoology*, **49**, 67–80.

Oyama, S. (1985). *The Ontogeny of Information: Developmental Systems and Evolution*. Cambridge: Cambridge University Press.

(1999). Locating development: Locating developmental systems. In *Conceptual Development: Piaget's Legacy*, ed. E.K. Scholnick, K. Nelson, S.A. Gelman, and P.H. Miller, pp. 185–208. Mahwah, NJ: Lawrence Erlbaum Associates.

(2000a). *Evolution's Eye: A Systems View of the Biology-Culture Divide*. Durham, NC: Duke University Press.

(2000b). *The Ontogeny of Information: Developmental Systems and Evolution*, rev. ed. Durham, NC: Duke University Press.

Oyama, S., Griffiths, P.E., and Gray, R., eds. (2001). *Cycles of Contingency: Developmental Systems and Evolution*. Cambridge, MA: MIT Press.

(2001). Introduction: What is developmental systems theory? In *Cycles of Contingency: Developmental Systems and Evolution*, ed. S. Oyama, P.E. Griffiths, and R. Gray, pp. 1–11. Cambridge, MA: MIT Press.

Patel, N.H. (1994). Developmental evolution: Insights from studies of insect segmentation. *Science*, **266**, 581–590.

Pennisi, E. (2000). Embryonic lens prompts eye development. *Science*, **289**, 522–523.

Peterson, K., and Sapienza, C. (1993). Imprinting the genome: Imprinted genes, imprinting genes, and a hypothesis for their interaction. *Annual Review of Genetics*, **27**, 7–31.

Petronis, A. (2001). Human morbid genetics revisited: Relevance of epigenetics. *Trends in Genetics*, **17**, 142–146.

Pigliucci, M. (2001). *Phenotypic Plasticity: Beyond Nature and Nurture*. Baltimore: Johns Hopkins University Press.

Pigliucci, M., and Schlichting, C.D. (1997). On the limits of quantitative genetics for the study of phenotypic evolution. *Acta Biotheoretica*, **45**, 143–160.

Pinto-Correia, C. (1997). *The Ovary of Eve: Egg and Sperm and Preformation*. Chicago: University of Chicago Press.

—— (1999). Strange tales of small men: Homunculi in reproduction. *Perspectives in Biology and Medicine*, **42**, 225–244.

Plotkin, H. (1994). *Darwin Machines and the Nature of Knowledge*. Toronto: Penguin Press.

Poerksen, U. (1995). *Plastic Words: The Tyranny of a Modular Language*, trans. J. Mason and D. Cayley. University Park: Pennsylvania State University Press.

Raff, R.A. (1996). *The Shape of Life: Genes, Development, and the Evolution of Animal Form*. Chicago: University of Chicago Press.

Rehmann-Sutter, C. (1996). Frankensteinian knowledge? *The Monist*, **79**, 264–279.

Riedl, R. (1977). A systems-analytical approach to macro-evolutionary phenomena. *Quarterly Review of Biology*, **52**, 351–370.

Rieppel, O. (2001). Turtles as hopeful monsters. *BioEssays*, **23**, 987–991.

Riggs, A.D., and Porter, T.N. (1996). Overview of epigenetic mechanisms. In *Epigenetic Mechanisms of Gene Regulation*, ed. V.E.A. Russo, R.A. Martienssen, and A.D. Riggs, pp. 29–45. Plainview, TX: Cold Spring Harbor Laboratory Press.

Robert, J.S. (2000a). Schizophrenia epigenesis? *Theoretical Medicine & Bioethics*, **21**, 191–215.

—— (2000b). Wild ontology: Elaborating environmental pragmatism. *Ethics & the Environment*, **5**, 191–209.

—— (2000c). Fastidious, foundational heresies. *Biology & Philosophy*, **15**, 133–145.

—— (2000d). Synthetic biology. *Studies in History and Philosophy of Science, Part C: Biological and Biomedical Sciences*, **31**, 599–614.

—— (2001a). Interpreting the homeobox: Metaphors of gene action and activation in development and evolution. *Evolution & Development*, **3**, 287–295.

—— (2001b). Genomes, hormones and health. *Literary Review of Canada*, **9.4**, 18–21.

—— (2002). How developmental is evolutionary developmental biology? *Biology & Philosophy*, **17**, 591–611.

—— (2003). Developmental systems and animal behaviour. *Biology & Philosophy*, **18**, 477–489.

—— (in preparation a). *Healthy Genomes, Healthy Folks?*.

—— (in preparation b). Integrating biological disciplines.

Robert, J.S., Hall, B.K., and Olson, W.M. (2001). Bridging the gap between developmental systems theory and evolutionary developmental biology. *BioEssays*, **23**, 954–962.

Roll-Hansen, N. (1984). E.S. Russell and J.H. Woodger: The failure of two twentieth-century opponents of mechanistic biology. *Journal of the History of Biology*, **17**, 399–428.

Rose, S. (1997). *Lifelines: Biology Beyond Determinism*. New York: Oxford University Press.

Rosenberg, A. (1997). Reductionism redux: Computing the embryo. *Biology & Philosophy*, **12**, 445–470.

Roskam, J.C., and Brakefield, P.M. (1999). Seasonal polyphenism in Bicyclus (*Lepi-doptera: Satyridae*) butterflies: Different climates need different cues. *Biological Journal of the Linnean Society*, **66**, 345–356.

Roux, W. (1894). The problems, methods, and scope of developmental mechanics, trans. W.M. Wheeler. In *Wood's Holl Biological Lectures for 1894*, pp. 149–190. Boston: Ginn & Company. Original edition, 1895.

Ruse, M. (1975). Woodger on genetics: A critical evaluation. *Acta Biotheoretica*, **24**, 1–13.

Russell, E.S. (1916). *Form and Function: A Contribution to the History of Animal Morphology*. London: John Murray.

(1930). *The Interpretation of Development and Heredity*. Oxford: Clarendon Press.

(1933). The limitations of analysis in biology. *Proceedings of the Aristotelian Society*, **33**, 147–158.

Russo, V.E.A., Martienssen, R.A., and Riggs, A.D., eds. (1996). *Epigenetic Mechanisms of Gene Regulation*. Cold Spring Harbour, NY: Cold Spring Harbor Laboratory Press.

Sander, K. (1986). The role of genes in ontogenesis – evolving concepts from 1883 to 1983 as perceived by an insect embryologist. In *A History of Embryology*, ed. T.J. Horder, J.A. Witkowski, and C.C. Wylie, pp. 363–395. Cambridge: Cambridge University Press.

Sapp, J. (1987). *Beyond the Gene: Cytoplasmic Inheritance and the Struggle for Authority in Genetics*. Oxford: Oxford University Press.

(1991). Concepts of organization: The leverage of ciliate protozoa. In *A Conceptual History of Modern Embryology*, ed. S.F. Gilbert, pp. 229–258. Baltimore: Johns Hopkins University Press.

Sarkar, S. (1996a). Biological information: A skeptical look at some central dogmas of molecular biology. In *The Philosophy and History of Molecular Biology: New Perspectives*, ed. S. Sarkar, pp. 187–231. Dordrecht: Kluwer.

(1996b). Decoding 'coding' – information and DNA. *BioScience*, **46**, 857–864.

(1998). *Genetics and Reductionism*. Cambridge: Cambridge University Press.

(1999). From the Reaktionsnorm to the adaptive norm: The norm of reaction, 1909–1960. *Biology & Philosophy*, **14**, 235–252.

Sarkar, S., and Robert, J.S. (2003). Editors' introduction [to a special issue on evolution and development]. *Biology & Philosophy*, **18**, 209–217.

Schaffner, K.F. (1998). Genes, behavior, and developmental emergentism: One process, indivisible? *Philosophy of Science*, **65**, 209–252.

Schank, J.C., and Wimsatt, W.C. (1986). Generative entrenchment and evolution. *PSA-1986: Philosophy of Science Association (Proceedings)*, Vol. 2, 33–60.

(2001). Evolvability, adaptation, and modularity. In *Thinking About Evolution: Historical, Philosophical, and Political Perspectives*, ed. R.S. Singh, C.B. Krimbas, D.B. Paul, and J. Beatty, pp. 322–335. Cambridge: Cambridge University Press.

Schlichting, C.D., and Pigliucci, M. (1998). *Phenotypic Evolution: A Reaction Norm Perspective*. Sunderland, MA: Sinauer Associates, Inc.

Schrödinger, E. (1944). *What Is Life?* Cambridge: Cambridge University Press.

Schwartz, J.H. (1999). *Sudden Origins: Fossils, Genes, and the Emergence of Species*. Toronto: John Wiley & Sons.

Bibliography

Scriver, C.R., and Waters, P.J. (1999). Monogenic traits are not simple: Lessons from phenylketonuria. *Trends in Genetics*, **15**, 267–272.

Shishkin, M.A. (1992). Evolution as a maintenance of ontogenetic stability. *Acta Zoologica Fennica*, **191**, 37–42.

Smith, K.C. (1992). The new problem of genetics: A response to Gifford. *Biology & Philosophy*, **7**, 431–452.

   (1993). Neo-Rationalism vs. Neo-Darwinism: Integrating development and evolution. *Biology & Philosophy*, **7**, 431–451.

Smith, K.K., and Schneider, R.A. (1998). Have gene knockouts caused evolutionary reversals in the mammalian first arch? *BioEssays*, **20**, 245–255.

Sober, E. (2000). Appendix I: The meaning of genetic causation. In A. Buchanan, D.W. Brock, N. Daniels, and D. Wikler, *From Chance to Choice: Genetics and Justice*, pp. 347–370. Cambridge: Cambridge University Press.

Solé, R.V., Salazar-Ciudad, I., and Newman, S.A. (2000). Gene network dynamics and the evolution of development. *Trends in Ecology and Evolution*, **15**, 479–480.

Spencer-Smith, R. (1994–1995). Reductionism and emergent properties. *Proceedings of the Aristotelian Society*, **95**, 113–129.

Sterelny, K. (2000a). The 'genetic program' program: A commentary on Maynard Smith on information in biology. *Philosophy of Science*, **67**, 195–201.

   (2000b). Development, evolution, and adaptation. *Philosophy of Science*, **67**(Supplement: Proceedings of the 1998 Biennial Meetings of the Philosophy of Science Association; Part II: Symposia Papers), S369–S387.

   (2001). Niche construction, developmental systems, and the extended replicator. In *Cycles of Contingency: Developmental Systems and Evolution*, ed. S. Oyama, P. Griffiths, and R. Gray, pp. 333–349. Cambridge, MA: MIT Press.

Sterelny, K., and Griffiths, P.E. (1999). *Sex and Death: An Introduction to Philosophy of Biology*. Chicago: University of Chicago Press.

Sterelny, K., and Kitcher, P. (1988). The return of the gene. *Journal of Philosophy*, **85**, 339–361.

Sterelny, K., Smith, K.C., and Dickison, M. (1996). The extended replicator. *Biology & Philosophy*, **11**, 377–403.

Stern, D.L. (2000). Evolutionary developmental biology and the problem of variation. *Evolution*, **54**, 1079–1091.

Strohman, R.C. (1993). Ancient genomes, wise bodies, unhealthy people: Limits of a genetic paradigm in biology and medicine. *Perspectives in Biology and Medicine*, **37**, 112–145.

   (1997). Profit margins and epistemology. *Nature Biotechnology*, **15**, 1224–1226.

Szathmáry, E. (1999). When the means do not justify the end. *Nature*, **399**, 745.

Thom, R. (1989). An inventory of Waddingtonian concepts. In *Theoretical Biology: Epigenetic and Evolutionary Order from Complex Systems*, ed. B. Goodwin and P. Saunders, pp. 1–7. Edinburgh: Edinburgh University Press.

Venter, J.C., Adams, M.D., and Myers, E.W. (2001). The sequence of the human genome. *Science*, **291**, 1304–1351.

Vinci, T., and Robert, J.S. (in preparation). Aristotle and modern genetics.

Vogel, G. (2000). A mile-high view of development. *Science*, **288**, 2119–2120.

Waddington, C.H. (1952). *The Epigenetics of Birds*. Cambridge: Cambridge University Press.

(1961). Genetic assimilation. *Advances in Genetics*, **10**, 257–293.

(1975). *The Evolution of an Evolutionist*. Ithaca, NY: Cornell University Press.

Wade, N. (2002). Scientist reveals genome secret: It's his. *New York Times* (27 April), available online at <http://www.nytimes.com/2002/04/27/science/27GENO.html>.

Wagner, G.P. (2000). What is the promise of developmental evolution? Part I: Why is developmental biology necessary to explain evolutionary innovations? *Journal of Experimental Zoology (Molecular and Developmental Evolution)*, **288**, 95–98.

(2001). What is the promise of developmental evolution? Part II: A causal explanation of evolutionary innovations may be impossible. *Journal of Experimental Zoology (Molecular and Developmental Evolution)*, **291**, 305–309.

Wagner, G.P., Laubichler, M., and Chiu, C.-H. (2000). Developmental evolution as a mechanistic science: The inference from developmental mechanisms to evolutionary processes. *American Zoologist*, **40**, 819–831.

Wahlsten, D., (1990). Insensitivity of the analysis of variance to heredity-environment interaction. *Behavioral and Brain Sciences*, **13**, 109–161.

Wahlsten, D., and Gottlieb, G. (1997). The invalid separation of effects of nature and nurture: Lessons from animal experimentation. In *Intelligence, Heredity, and Environment*, ed. R.J. Sternberg and E. Grigorenko, pp. 163–192. Cambridge: Cambridge University Press.

Wallace, B. (1986). Can embryologists contribute to an understanding of evolutionary mechanisms? In *Integrating Scientific Disciplines*, ed. W. Bechtel, pp. 149–163. Dordrecht: Martinus Nijhoff.

van der Weele, C. (1999). *Images of Development: Environmental Causes in Ontogeny*. Albany: State University of New York Press.

Weismann, A. (1893). *The Germ Plasm*, trans. W. Newton Parker and H. Rönnfeldt. New York: Charles Scribner's Sons.

Weiss, K.M., and Fullerton, S.M. (2000). Phenogenetic drift and the evolution of genotype-phenotype relationships. *Theoretical Population Biology*, **57**, 187–195.

West, M.J., and King, A.P. (1987). Settling nature and nurture into an ontogenetic niche. *Developmental Psychobiology*, **20**, 549–562.

West-Eberhard, M.-J. (1998). Evolution in the light of developmental and cell biology, and vice versa. *Proceedings of the National Academy of Sciences U.S.A.*, **95**, 8417–8419.

Whitman, C.O. (1894). Evolution and epigenesis. In *Wood's Holl Biological Lectures for 1894*, pp. 205–224. Boston: Ginn & Company.

Williams, G.C. (1992). *Natural Selection: Domains, Levels and Challenges*. Oxford: Oxford University Press.

Wilson, E.B. (1925). *The Cell in Development and Inheritance*, 3rd ed. New York: Macmillan & Co.

Wilson, E.O. (1975). *Sociobiology: The New Synthesis*. Cambridge, MA: Harvard University Press.

Wimsatt, W.C. (1980). Reductionistic research strategies and their biases in the units of selection controversy. In *Scientific Discovery, Vol. 2: Case Studies*, ed. T. Nickles, pp. 213–259. Dordrecht: D. Reidel.

(1986a). Forms of aggregativity. In *Human Nature and Natural Knowledge*, ed. A. Donagan, N. Perovich, and M. Wedin, pp. 259–293. Dordrecht: D. Reidel.

(1986b). Developmental constraints, generative entrenchment, and the innate-acquired distinction. In *Integrating Scientific Disciplines*, ed. W. Bechtel, pp. 185–208. Dordrecht: Martinus Nijhoff.

(1986c). Heuristics and the study of human behavior. In *Metatheory in Social Science: Pluralisms and Subjectivities*, ed. D.W. Fiske and R.A. Shweder, pp. 293–314. Chicago: University of Chicago Press.

(1987). False models as means to truer theories. In *Neutral Models in Biology*, ed. M.H. Nitecki and A. Hoffman, pp. 23–55. Oxford: Oxford University Press.

(1997). Aggregativity: Reductive heuristics for finding emergence. *Philosophy of Science*, **64** (Supplement: Proceedings of the 1996 Biennial Meetings of the Philosophy of Science Association; Part II: Symposia Papers), S372–S384.

(1999). Genes, memes and cultural heredity. *Biology & Philosophy*, **14**, 279–310.

(2001). Generative entrenchment and the developmental systems approach to evolutionary processes. In *Cycles of Contingency*, ed. S. Oyama, P.E. Griffiths, and R. Gray, pp. 219–237. Cambridge, MA: MIT Press.

Wimsatt, W.C., and Schank, J.C., (1988). Two constraints on the evolution of complex adaptations and the means for their avoidance. In *Evolutionary Progress*, ed. M. Nitecki, pp. 231–273. Chicago: University of Chicago Press.

Winther, R.G. (2001). Varieties of modules: kinds, levels, origins and behaviors. *Journal of Experimental Zoology (Molecular and Developmental Evolution)*, **291**, 116–129.

Wolf, U. (1995). The genetic contribution to the phenotype. *Human Genetics*, **95**, 127–148.

Wolffe, A.P. (1998). Introduction. *Epigenetics: Novartis Foundation Symposium*, **214**, 1–5.

Wolpert, L. (1991). *The Triumph of the Embryo*. Oxford: Oxford University Press.

(1994). Do we understand development? *Science*, **266**, 571–572.

(1995). Development: Is the egg computable, or could we generate an angel or a dinosaur? In *What Is Life? The Next Fifty Years: Speculations on the Future of Biology*, ed. M.P. Murphy and L.A.J. O'Neill, pp. 57–66. Cambridge: Cambridge University Press.

Woodger, J.H. (1952). *Biology and Language*. Cambridge: Cambridge University Press.

(1959). Studies in the foundation of genetics. In *The Axiomatic Method with Special Reference to Geometry and Physics*, ed. L. Henkin, P. Suppes, and A. Tarski, pp. 408–428. Amsterdam: North-Holland Publishing Co.

Yamamoto, Y., and Jeffery, W.R. (2000). Central role for the lens in cave fish eye degeneration. *Science*, **289**, 631–633.

Index

157